Innovative Processing and Synthesis of Ceramics, Glasses and Composites VIII

T0329154

Technical Resources

Journal of the American Ceramic Society

www.ceramicjournal.org

With the highest impact factor of any ceramics-specific journal, the *Journal of the American Ceramic Society* is the world's leading source of published research in ceramics and related materials sciences.

Contents include ceramic processing science; electric and dielectic properties; mechanical, thermal and chemical properties; microstructure and phase equilibria; and much more.

Journal of the American Ceramic Society is abstracted/indexed in Chemical Abstracts, Ceramic Abstracts, Cambridge Scientific, ISI's Web of Science, Science Citation Index, Chemistry Citation Index, Materials Science Citation Index, Reaction Citation Index, Current Contents/ Physical, Chemical and Earth Sciences, Current Contents/Engineering, Computing and Technology, plus more.

View abstracts of all content from 1997 through the current issue at no charge at www.ceramicjournal.org. Subscribers receive full-text access to online content.

Published monthly in print and online. Annual subscription runs from January through December. ISSN 0002-7820

International Journal of Applied Ceramic Technology

www.ceramics.org/act

Launched in January 2004, *International Journal of Applied Ceramic Technology* is a must read for engineers, scientists,and companies using or exploring the use of engineered ceramics in product and commercial applications.

Led by an editorial board of experts from industry, government and universities, *International Journal of Applied Ceramic Technology* is a peer-reviewed publication that provides the latest information on fuel cells, nanotechnology, ceramic armor, thermal and environmental barrier coatings, functional materials, ceramic matrix composites, biomaterials, and other cutting-edge topics.

Go to www.ceramics.org/act to see the current issue's table of contents listing state-of-the-art coverage of important topics by internationally recognized leaders.

Published quarterly. Annual subscription runs from January through December. ISSN 1546-542X

American Ceramic Society Bulletin

www.ceramicbulletin.org

The *American Ceramic Society Bulletin*, is a must-read publication devoted to current and emerging developments in materials, manufacturing processes, instrumentation, equipment, and systems impacting the global ceramics and glass industries.

The *Bulletin* is written primarily for key specifiers of products and services: researchers, engineers, other technical personnel and corporate managers involved in the research, development and manufacture of ceramic and glass products. Membership in The American Ceramic Society includes a subscription to the *Bulletin*, including online access.

Published monthly in print and online, the December issue includes the annual *ceramicSOURCE* company directory and buyer's guide. ISSN 0002-7812

Ceramic Engineering and Science Proceedings (CESP)

www.ceramics.org/cesp

Practical and effective solutions for manufacturing and processing issues are offered by industry experts. CESP includes five issues per year: Glass Problems, Whitewares & Materials, Advanced Ceramics and Composites, Porcelain Enamel. Annual subscription runs from January to December. ISSN 0196-6219

ACerS-NIST Phase Equilibria Diagrams CD-ROM Database Version 3.0

www.ceramics.org/phasecd

The ACerS-NIST Phase Equilibria Diagrams CD-ROM Database Version 3.0 contains more than 19,000 diagrams previously published in 20 phase volumes produced as part of the ACerS-NIST Phase Equilibria Diagrams Program: Volumes I through XIII; Annuals 91, 92 and 93; High Tc Superconductors I & II; Zirconium & Zirconia Systems; and Electronic Ceramics I. The CD-ROM includes full commentaries and interactive capabilities.

Innovative Processing and Synthesis of Ceramics, Glasses and Composites VIII

Ceramic Transactions Volume 166

Proceedings of the 106th Annual Meeting of The American Ceramic Society, Indianapolis, Indiana, USA (2004)

Editors
Narottam P. Bansal
J. P. Singh
Hartmut Schneider

Published by
The American Ceramic Society
PO Box 6136
Westerville, Ohio 43086-6136
www.ceramics.org

Innovative Processing and Synthesis of Ceramics, Glasses and Composites VIII

Contents

Plasma Synthesis

Composites

Thin Films

Porous Ceramics

Kinetics and Mechanism

Computational Modeling and Analysis

Preface

An international symposium, "Innovative Processing and Synthesis of Ceramics, Glasses, and Composites," was held during the 106th Annual Meeting of The American Ceramic Society in Indianapolis, IN, April 18-21, 2004. This symposium provided a forum for scientists, engineers, and technologists to discuss and exchange state-of-the-art ideas, information, and technology on advanced methods and approaches for processing and synthesis of ceramics, glasses, and composites. A total of 75 papers, including six invited talks, were presented in the form of oral and poster presentations indicating continued interest in the scientifically and technologically important field of ceramic processing. Authors from 14 countries (Belarus, Belgium, Germany, Iran, Japan, Korea, Mexico, Portugal, Russia, Spain, Sweden, Taiwan, United Kingdom, and the United States) participated. The speakers represented universities, industries, and government research laboratories.

These proceedings contain contributions on various aspects of synthesis and processing of ceramics, glasses, and composites that were discussed at the symposium. Nineteen papers describing the latest developments in the areas of combustion synthesis, microwave processing, sol-gel synthesis, colloidal processing, pre-ceramic polymer processing, and plasma synthesis for fabrication of powders, electronic ceramics, composites, thin films, ceramic inks, and porous ceramics, etc. are included in this volume. Kinetics and mechanism of processes as well as computational modeling and analysis are also discussed. Each manuscript was peer-reviewed using The American Ceramic Society review process.

The editors wish to extend their gratitude and appreciation to all the authors for their cooperation and contributions, to all the participants and session chairs for their time and efforts, and to all the reviewers for their useful comments and suggestions. Financial support from The American Ceramic Society is gratefully acknowledged. Thanks are due to the staff of the Meetings and Publications departments of The American Ceramic Society for their invaluable assistance. Special thanks go to Greg Geiger of The American Ceramic Society for efficiently coordinating the on-site review of the manuscripts. His cheerful assistance and cooperation throughout the production process of this volume was of great help in making this task much easier and is very much appreciated.

It is our earnest hope that this volume will serve as a valuable reference for the researchers as well as the technologists interested in innovative approaches for synthesis and processing of ceramics, glasses, and composites.

Narottam P. Bansal
J. P. Singh
Hartmut Schneider

Combustion Synthesis

COMBUSTION SYNTHESIS (SHS) OF BN/AlN CERAMIC COMPOSITE POWDERS

Hayk H. Khachatryan, Mchitar A. Hobosyan, and Suren L. Kharatyan
A.B. Nalbandyn Institute of Chemical Physics, National Academy of Science
Yerevan 375044, Republic of Armenia

Jan A. Puszynski
South Dakota School of Mines and Technology
Rapid City, SD 57701

ABSTRACT
    Combustion synthesis is a very energy efficient and effective method of producing ceramic composite powders and multi-component solid solutions in a single reaction step. In addition, this process, also called self-propagating high temperature synthesis (SHS), leads to the formation of products with grater homogeneity than other commercially available synthesis methods. In this study, BN-AlN composite powders were synthesized in a self-sustaining regime. Aluminum diboride (AlB_2) reactant together with AlN and/or BN diluents were mixed and ignited at elevated nitrogen pressures (> 2 MPa). This paper addresses a role of diluents (AlN or BN) and their concentrations as well as nitrogen pressure on combustion front velocity, combustion temperature, overall conversion, and product phase composition. The reaction mechanism and thermodynamic analyses of Al-B-N combustion system are also presented.

INTRODUCTION
    There is a continuous need for new materials with improved mechanical, thermal, and corrosion resistance properties. Many nonoxide ceramic materials and ceramic-ceramic composites have exceeding properties of metals, especially in high temperature and corrosive applications. Aluminum nitride has proven to be an excellent material exhibiting high thermal conductivity and resistance to molten aluminum. However, this material does not posses superior thermal shock resistance. An addition of hexagonal boron nitride to aluminum nitride or other non-oxide ceramics significantly improves thermal shock resistance of resulting composites. Wear resistance, corrosion resistance, and machinability of such composites are also improved [1-3]. Aluminum nitride - boron nitride composites along with their good machinability, are characterized by high thermal conductivity, high electrical resistivity, and good dielectric properties.
    Up to now, commercial boron nitride-based composites have been prepared by three techniques: 1) hot pressing of premixed powders in nitrogen atmosphere, 2) displacement reactions followed by sintering, and 3) chemical vapor deposition.
    In this study another synthesis technique has been considered and explored. The combustion synthesis is in many cases very economical method of producing non-oxide ceramics [4]. This technique allows synthesizing nonstochiometric composites or solid solutions, which are difficult to obtain by the traditional furnace technology. The principle of this method is to form desired products by igniting strongly exothermic reaction between condensed-phase reactants ($Ti+2B \rightarrow TiB_2$) or solid and gaseous reactants (e.g. $Al+1/2N_2 \rightarrow AlN$). These SHS reactions, once ignited, generate high temperatures (typically >2000°C) and the combustion front propagates with the velocity in the range of millimeters or centimeters per second. In addition, combustion synthesized powders may exhibit higher homogeneity and better sinterability [5].

The versatility of combustion reaction allows synthesis of different materials with relativity high throughput.

The objective of this study was threefold: 1) conduct thermodynamic analysis of four reacting systems to identify the best synthesis condition; 2) investigate the formation of BN-AlN composites by SHS technique, and 3) characterize combustion synthesized products.

THERMODYNAMIC CONSIDERATIONS

The following four different reactions were considered in this study.

$$2B + nAlN + N_2 \rightarrow 2BN + nAlN$$
$$2Al + nBN + N_2 \rightarrow 2AlN + nBN$$
$$AlB_2 + nAlN + 1.5N_2 \rightarrow (n+1)AlN + 2BN$$
$$AlB_2 + nBN + 1.5N_2 \rightarrow AlN + (2+n)BN$$

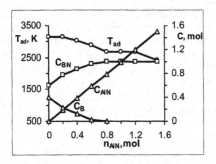
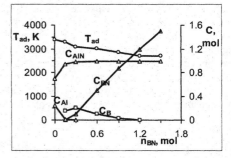

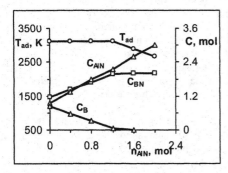
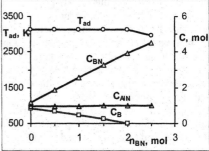

Based on thermodynamic calculations done by using thermodynamic software THERMO from ISMAN Chernogolovka, Russia, it was found that free boron is still present in the product at lower nitrogen pressures and low degree of reactant dilution ($P_{N2} < 2$ MPa). The results were confirmed by using FACT and HSC thermodynamic software packages. Increased concentration of AlN or BN in the initial mixture leads to the significant reduction of free boron in the final product. Therefore, in order to find compromise between high and low pressures and adequate level of dilution, the nitrogen pressure of 2 MPa was chosen. Figure 1 shows the effect of AlN and BN dilution on equilibrium composition of products and adiabatic temperature for four above described reactions. In all considered reactions free boron was found below certain level of the dilution with BN or AlN (see Figure 1). Thermodynamic analysis of AlB_2-AlN-N_2 system indicates that the nitridation of AlB_2 occurs in three stages (Figure 2): a) formation of aluminum nitride and aluminum boride, AlB_{12}; b) further nitridation of AlB_{12} resulting in the formation of AlN and partial nitridation of boron; c) final nitridation of boron.

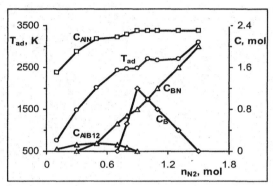

The conclusions from this thermodynamic analysis are supported by the experimental results presented by Zhang et al [6]. They showed that nitridation of AlB_2 at 1 atm nitrogen pressure follows a consecutive reaction scheme. First AlN and $Al_{1.61}B_{22}$ are formed followed by the formation of AlN-BN composite.

EXPERIMENTAL

All combustion synthesis experiments were performed in a high-pressure reactors. The experimental setup used in this study was described elsewhere [7].

The following reactant powders were used in the experiments: 1) AlB_2 with $d_{50}=4.5\mu m$ and $d_{90}=24\mu m$, 2) BN with different average particle size, $10\mu m$, $25\mu m$, $50\mu m$, $65\mu m$, $70\mu m$, $80\mu m$, $125\mu m$, and $>125\mu m$, and 3) AlN $d_{50}=4.9\mu m$, and $d_{90}=29\mu m$. Nitrogen gas with 99.97% purity was used in all experiments.

The reactant powders were mixed in a ball mill for 8 hrs. After that cylindrical pellets with a diameter of 35mm and height of 40-50mm were pressed. An average bulk density of pressed pellets was between 0.5-$0.6g/cm^3$. The pressed pellets were placed in the high pressure reactor, which was evacuated, purged, and pressurized with nitrogen prior to the ignition. The ignition was done by a resistively heated tungsten coil at 2 MPa nitrogen pressure. Combustion temperature, T_c, and average combustion front velocities, U_c, were measured by tungsten - rhenium thermocouples with 0.1mm diameter. These thermocouples were placed at different axial positions in the pellet. After combustion process was completed, the product was grinded and milled in the ball or vibratory mill.

RESULTS AND DISCUSSION
AlB$_2$-BN-N$_2$ Reacting System

Both AlB$_2$ and BN reactants were synthesized in house using a combustion technique. SEM photographs of both powders used in this study are shown in Figures 3 and 4. As can be seen from these photographs aluminum diboride powder has more like spherical shape, when hexagonal boron nitride is in a form of plateletts.

Combustion synthesis experiments, conducted under 2 MPa nitrogen pressure, showed that the maximum combustion temperature (~2200°C) is rather not affected by the dilution with boron nitride in a wide range of concentrations (0-65 wt%). Interestingly above the higher level of dilution (>65 wt%), the nitridation reaction is not self-sustaining. In contrary, the combustion front propagation velocity is significantly affected by the level of BN dilution (see Figure 5). Up to 40 wt%BN this velocity increases with increasing degree of dilution. At low BN concentrations, the aluminum diboride powder decomposes above 2100K and the resulting molten boron and aluminum tend to coalescence prior to the nitridation. Examination of pellet's microstructure has clearly indicated melting of intermediate Al and B products. The increase of the content of BN diluent reduced the effect of intermediate products coalescence and their sintering. In addition, the dilution enhances a nitridation process by providing additional surface area of a refractory materials carrier. At higher dilution ratios (above 40 wt%) the combustion front propagation velocity decreases with the increasing concentration of the diluent. This decrease might be caused by increased resistance to a nitrogen transport within the pellet. The overall conversion, η, of AlB$_2$ calculated from the total weight change is always less then 100%. This is caused by incomplete reaction close to the pellet's surface. Significant heat losses from the outer surface of the pellet are responsible for a low level of conversion in that region.

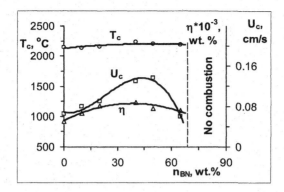

In Figure 6, dynamic temperature profiles measured inside pellets with two different initial compositions are shown. It can be clearly seen that the rate of cooling in the pellet with higher concentration of BN is less pronounced then that observed during the combustion of the pellet with the lower BN concentration. This lower rate of cooling indicates that nitridation reaction continues after the combustion front has passed. This observation provides additional evidence that the infiltration of nitrogen into the pellet plays a significant role.

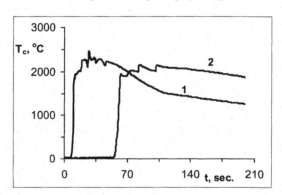

The effect of an average boron nitride particle size on key combustion characteristics was investigated as well. Figure 7 shows the variation of maximum combustion temperature, overall pellet conversion, and combustion front velocity as the function of an average particle size of BN diluent. It can be clearly seen that maximum combustion temperature is not affected by the size of diluent. It is interesting to note that the overall pellet's conversion and the combustion front propagation velocity are increasing with the average particle size of diluent. These results also support the strong effect of nitrogen transport limitation within the pellet.

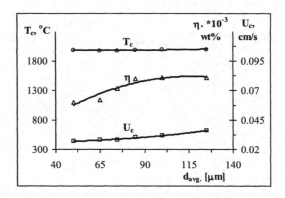

X-ray diffraction pattern of the combustion synthesized AlN-BN composite is shown in Figure 8. Only two, BN and AlN phases were detected. Specific surface area analysis (BET) of ball milled products showed specific surface areas of combustion synthesized powders to be between 8 and 24 m^2/g.

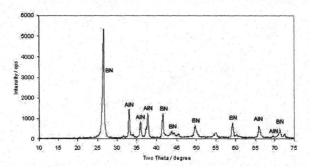

The effect of nitrogen gas pressure on key combustion characteristics in AlB_2-65 wt%BN system is shown in Figure 9. It was observed that this reaction system is not self-sustaining below 1.5 MPa. It was found that the nitrogen pressure has a pronounced effect on maximum combustion temperature, overall pellet conversion, and combustion front propagation velocity.

AlB_2-AlN-N_2 Reacting System

Figure 10 shows the effect of AlN concentration in the initial mixture on combustion characteristics of AlB_2-AlN system at 2 MPa nitrogen pressure. The propagation characteristics are very similar to those observed in AlB_2-BN-N_2 system (see Figure 5).

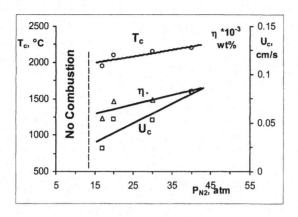

The effect of increasing concentration of AlN was less pronounced than in the case of boron nitride and the propagation limit was wider (up to 80 wt%). Self-sustaining character of the nitridation reaction in the presence of AlN was observed at much lower nitrogen pressure (P_{N2}=0.4 MPa, 75 wt %AlN) compare to the case when BN diluent (P_{N2}>1.5 MPa and 65 wt%BN) was used (see Figures 9 and 11).

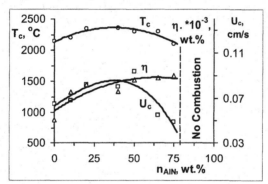

This behavior could be explained by different morphology of AlN (spherical shape) in contrary to BN which was in the form of plateletts. The overall gas filtration resistance in the pellets with AlN diluent could be lower, however additional permeation studies are needed to confirm this hypothesis.

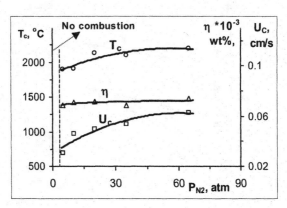

Figure 11. Combustion parameters of AlB$_2$-75 wt%AlN-
N$_2$ system as the function of nitrogen pressure.

CONCLUSIONS

1. It was found that AlN-BN composites with a high specific surface area (8-24m^2/g) and with the wide range of compositions can be synthesized by the combustion synthesis technique.

2. Stable combustion propagation fronts in AlB$_2$-AlN or AlB$_2$-BN systems were observed at nitrogen pressures above 2 MPa.

3. It was shown that the nitrogen pressure has a significant effect on the overall conversion. It was also demonstrated that AlN - BN composites can be synthesized at 2 MPa nitrogen pressure with 85-95% conversion within the certain ranges of BN or AlN average particle sizes and degrees of dilution.

ACKNOWLEDGMENT
The authors would like to acknowledge the financial support from the Civilian Research Development Foundation (Grant № NFSAT CH 123-02/CRDF 12047).

REFERENCES

1. K.S. Mazdiyasni, R. Ruh, E.E. Hermes, "Phase Characterization and Properties of AlN-BN Composites". J. Am. Ceram. Bull., 64 [8] 1149-54 (1985).

2. S. Du, Zh. Liu, L. Gao, F. Li, L. Li, "Dielectric Properties of AlN-BN Composite Ceramics", J. Am. Ceram. Bull., 69-73 (1997).

3. T.D. Xiao, K.E. Gonsalves, P.R. Stritt, "Synthesis of Aluminum Nitride/Boron Nitride Composite Materials", J. Am. Ceram. Soc., 76 [4] 987-92 (1993).

4. A.G. Merzhanov, "Theory and Practice of SHS: Worldwide State of the Art and the Newest Results," International Journal of Self-Propagating High-Temperature Synthesis, 2 [2] 113-158 (1993).

5. J. Lis, S. Majorowski, J.A. Puszynski and V. Hlavacek; "Densification of Combustion Synthesized Silicon Nitride", Ceram. Bull. 70 [2], 244, (1991).

6. G.J. Zhang, J.F. Yang, M. Ando, T. Ohji, "Reaction Synthesis of Aluminum Nitride-Boron Nitride Composite Based on the Nitridation of Aluminum Boride", J. Am. Ceram. Soc., 85 [12] 2938-44 (2002).

7. L.S. Abovyan, H.H. Nersisyan, S.L. Kharatyan, R, Saiu, R. Orru, G. Cao and D. Zedda. "Synthesis of Alumina Silicon Carbide Composites by Chemically Activated Self-Propagating Reactions", Ceramics International, 27 [2] 163-169 (2001).

COMBUSTION SYNTHESIS OF CERAMIC COMPOSITES AND SOLID SOLUTIONS FROM NANOREACTANTS

Jan A. Puszynski
South Dakota School of Mines & Tech.
501 East Saint Joseph Street
Rapid City, SD 57701

Shivanee R. Dargar
South Dakota School of Mines & Tech.
501 East Saint Joseph Street
Rapid City, SD 57701

Berthold E. Liebig
South Dakota School of Mines & Tech.
501 East Saint Joseph Street
Rapid City, SD 57701

ABSTRACT

Submicron β-sialon powders with $z=1$ and $z=3$ were synthesized using combustion technique at high nitrogen pressures ($P_{N2} = 4$ MPa). These powders were synthesized by direct nitridation of aluminum and silicon nanoreactants in the presence of nano-size alumina and the final product as a diluent. In addition, in-situ formed titanium/nickel aluminides-alumina composites were densified by a direct application of uniaxial pressure during the reaction. As a result dense composites with ultrafine microstructures were obtained.

INTRODUCTION

Combustion synthesis, also called self-propagating high temperature synthesis (SHS), has been proven to be an economical method of synthesizing refractory complex solid solutions or composite powders. This technique utilizes a strongly exothermic character of solid-solid or gas-solid noncatalytic reactions [1-6]. The basic advantage of the SHS reaction is its self-sustaining feature. Once the SHS reaction is initiated, a combustion front is formed, which propagates converting reactants to products within seconds compare to hours for conventional furnace-based synthesis methods. Additional advantages of the SHS technique are higher purity of products by self-purification in the combustion zone and the potential of in-situ densification of synthesized products. Due to a relatively short exposure time at high temperatures usually products with smaller grain sizes are formed. During the combustion synthesis high temperatures are generated and in some cases they may exceed 3000K [6]. When external pressure is applied to hot combustion products a partial or complete densification may be accomplished depending on the level of applied pressure or fraction of liquid product(s) formed during the reaction [7-11].

During the past three decades, the majority of research in the area of combustion synthesis was done using micron-size reactants. Recently, several attempts have been made to reduce a diffusion length between solid reactants by ultimate dry mixing of micron-size reactants in high-energy mills [12-15]. This approach of pre-alloying is very interesting from scientific point of view, however a potential of fire hazard might prohibit scale-up of this process.

In this study, a different approach was undertaken. Instead of intensive mixing/milling of solid reactants, nanosized reactants have been explored both in gasless and gas-solid SHS reactions. The specific objectives of this research studies were twofold: i) synthesis of sialon powders from aluminum, silicon, and alumina nanopowders at elevated nitrogen pressures, and

ii) in-situ combustion synthesis and densification of reacting system consisting of a mixture of aluminum, nickel, and titanium dioxide nanopowders.

EXPERIMENTAL PROCEDURE
Two different types of SHS reactions were investigated in this study:
1. Direct nitridation of aluminum and silicon nanopowders in the presence of nanosized alumina at elevated nitrogen pressures, with the specific emphasis on the formation of sialon (z=1 and z=3) powders. The stoichiometric equations for the sialons, with z=1 and z=3 are:

$$15 \; Si + Al + Al_2O_3 \xrightarrow{\;\;N_2\;\;} 3 \; Si_5AlON_7 \quad (z=1)$$

$$3 \; Si + Al + Al_2O_3 \xrightarrow{\;\;N_2\;\;} Si_3Al_3O_3N_5 \quad (z=3)$$

2. Gasless reaction between aluminum/alumina and titanium dioxide nanopowders in the absence or presence of nickel nano-size reactant, followed by the uniaxial densification.

$$7 \; (Al/nAl_2O_3) + 3 \; TiO_2 \longrightarrow \text{Titanium aluminides} + (2+7n) \; Al_2O_3$$

$$8 \; (Al/nAl_2O_3) + 3 \; TiO_2 + Ni \longrightarrow \text{Titanium/nickel aluminides} + (2+7n) \; Al_2O_3$$

The following reactants were used in the above shown combustion reactions: i) aluminum from Technanogy Inc. (d_{avg}=50 nm, BET-SSA=37.10 m^2/g); ii) titanium dioxide from the Degussa Corporation (d_{avg}=40 nm); iii) silicon from Elkem Co. This micron-size powder was milled in the attritor for 16 hrs. BET surface area of attrition milled silicon powder was above 100 m^2/g; iv) micron-size aluminum from Valimet Inc. (d_{avg}=3.3 μm); v) α- alumina from Nanomat Inc. (d_{avg}=120 nm); vi) nickel from Argonide Co. (d_{avg}=72 nm).

Silicon, aluminum, and alumina nanoreactants were ultrasonically mixed in hexane or ethanol. After evaporation of the solvent, the reactant mixture was ignited in a high-pressure reactor. A detailed description of the experimental setup used in high-pressure nitridation experiments is described elsewhere [16].

Similarly stoichiometric amount of aluminum/alumina, titanium dioxide, and nickel nanoreactants were wet mixed in hexane using an ultrasonic bath. The wet mixing technique consists of several steps including dispersing of reactant mixture in hexane, quick draining on the plate to avoid stratification, drying and subsequent removing of powder mixture from drying plate. The mixture was placed inside Inconel die and uniaxial pressure of 100 MPa was applied to the upper punch. All experiments were carried out in the argon atmosphere. Graphite foil was placed along the die cavity, to prevent any reaction between nanoreactants and the Inconel die. The uniaxial pressure was controlled over the entire heating process. The required soak temperature was held for 1 hr. The furnace was then cooled and the pressed pellet was removed for further characterization. The experimental setup for in-situ combustion synthesis and densification of composites from nano-size powders is shown in Figure 1.

For comparison, a similar synthesis and densification process was done on the reacting system consisting of micron-size aluminum reactant. Dry mixing of that mixture was done in a rotary mixer for 3 hrs.

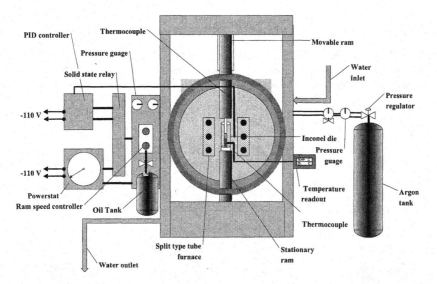

Figure 1. Schematic of in-situ combustion synthesis and densification of composites from nano-size powders.

RESULTS AND DISCUSSION

Mixing of Nano-size Reactants

One of the most important parameters affecting the quality of SHS generating powders is initial uniformity of the reactant mixture, especially when more than one solid reactant is involved. The preparation of a mixture consisting of several different nanopowders is even more challenging. Mixing intensity, surface properties of the powders, their interaction with a liquid (in most cases non-aqueous) and dispersants are the key factors affecting the final uniformity of binary or ternary mixtures. In this research a more comprehensive approach was undertaken to determine a completeness and final uniformity of binary aluminum-titanium dioxide mixtures. Both nanopowders were mixed in an equimolar ratio in anhydrous ethanol or hexane with 0.1 wt% of sodium dioctyl sulfosuccinate as a dispersant.

The slurry was placed in an ultrasonic bath where the final mixing took place for 30 minutes. The mixed samples were dried and analyzed using EDX-microprobe with a line scanning capability. The results from those X-ray analyses were used to determine the mixing index, A_{KL}, which was introduced by Kramers and Lacey [17]. This mixing index is defined as: $A_{K,L} = (\sigma_o - \sigma_a) / (\sigma_o - \sigma_R)$, where: σ_o=standard deviation of the initial mixture, σ_a=standard deviation of the actual mixture and σ_R=standard deviation of a random mixture. The standard deviation of a random mixture, σ_R, can be estimated as: $\sigma_R = [C_1 (1-C_1) /n]^{1/2}$, where: C_1 is fractional concentration of a key component in the mixture, and n is a sample size. The standard deviation of the initial mixture, σ_o, can be estimated as: $\sigma_o = [C_1/ (1-C_1)]^{1/2}$. The standard deviation of the actual mixture is: $\sigma_a = [\sum_{i=1}^{N}(C_i - C_{avg})^2 / (N-1)]^{1/2}$, where, C_i is local fraction of the key component and N is a number of analyzed samples. Figure 2 shows results of analyses of four aluminum-titanium dioxide samples mixed under different conditions. It can be clearly seen that the wet mixing in ethanol with surfactant or hexane with and without surfactant results in the similar values of A_{KL}, 0.950, 0.955 and 0.945 respectively. For comparison, dry mixing of this binary system resulted in a very low value of Lacey and Kramer index, A_{KL}=0.536.

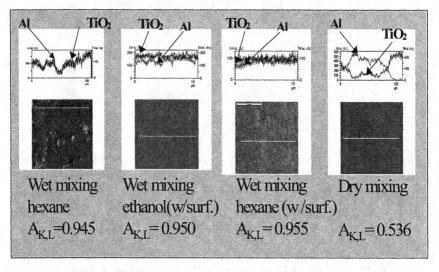

Figure 2. Comparison of mixing index after Lacey and Kramer for different mixing techniques.

Combustion Synthesis of Sialons

Two different β-sialon solid solutions with z=1 and z=3 were synthesized from aluminum, silicon, and alumina nanopowders at elevated nitrogen pressures (P_{N2} = 4 MPa). The nano-size reactants were mixed ultrasonically in anhydrous ethanol using the above-described mixing procedure. After drying, the mixture was placed inside a high-pressure reactor and ignited using a resistively heated molybdenum wire. The reaction products were characterized using X-ray diffraction and SEM. Figures 3 and 4 shows the X-ray diffractograms of combustion synthesized β-sialons, with z=1 and z=3, from nanoreactants. For the comparison, the X-ray diffractograms of β-sialons, with similar ultimate compositions formed from micron-size reactants are included. The corresponding X-ray diffractograms are matched quite well but X-ray peaks obtained from sialons formed from micron-size reactants are sharper indicating higher degree of crystallinity in the final product.

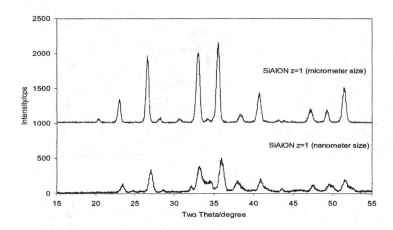

Figure 3. X-ray diffractogram of micron-size (above) and nano-size (below) β-sialon (z=1).

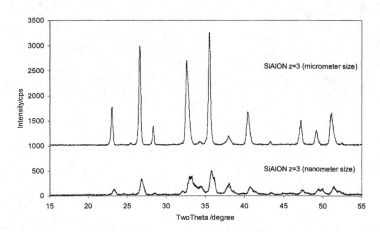

Figure 4. X-ray diffractogram of micron-size (above) and nano-size (below) β-sialon (z=3).

SEM photographs of combustion synthesized β-sialons, with z=1 and z=3, from nanopowders are shown in Figures 5 and 6 respectively. It is clearly seen that in both cases submicron powders were formed. However, finer powder was obtained in the case of β-sialon with z=1. SEM photographs of β-sialon with z=3 indicate more pronounced formation of whisker-like product. The evaluation of the effect of particle size of silicon, and aluminum as well as degree of dilution with the final product on propagation characteristics, powder morphology, and average particle size is currently underway and will be presented in a subsequent publication.

Figure 5. SEM image of nano-size β-sialon (z=1) (a) 10,000X (b) 5,000X.

Figure 6. SEM image of nano-size β-sialon (z=3) (a) 10,000X (b) with fibers 5,000X.

In-situ Combustion Synthesis and Densification of Composites

Aluminum nanopowders are protected with a thin alumina layer, and therefore they are not pyrophoric. However, the mass fraction of alumina is substantial and varies depending on the average particle size of the aluminum powder. For powders having average particle size of 50 nm, the alumina content is in a range between 25-40 wt%. TEM photograph of a commercial powder is shown in Figure 7. In the initial stage of this research, a densification of nano-size aluminum was carried out at temperatures between 450 and 600°C and uniaxial pressures of 50-100 MPa. Figure 8 shows the SEM photograph of an aluminum sample densified at 600°C and 100 MPa pressure. It can be clearly seen that grain size of densified Al-Al$_2$O$_3$ composite remains in the range of 60-100 nm. The SEM photograph also shows that the pellet has not been fully densified. The measured density by the Archimedes principle of densified Al-Al$_2$O$_3$ composites were 2.70-2.89 g/cc which corresponds to approximately 92-96 wt% of theoretical density.

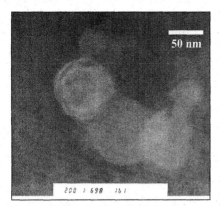

Figure 7. TEM photograph of aluminum nanopowders from Technanogy Inc.

Figure 8. SEM image of densified aluminum nanopowders at temperature of 600°C and pressure of 100 MPa.

The main focus of this investigation was on simultaneous combustion synthesis and uniaxial densification of products derived from aluminum, nickel, and titanium dioxide nanopowders. The combustion reaction was initiated by heating of the reactants mixture under constant uniaxial pressure of 100 MPa. The reacting system was kept at 940°C for 1 hr. Figure 9 shows the microstructure of the composite material derived from nanopowders, whereas Figure 10 shows SEM photograph of the composite formed from micron-size aluminum reactant and nano-size titanium dioxide having identical initial compositions.

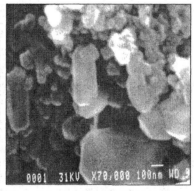

Figure 9. SEM image of titanium aluminides-alumina composite synthesized by nano-size reactants

Figure 10. SEM image of titanium aluminides-alumina composite synthesized by micron reactants.

Surprisingly, the X-ray diffraction patterns have shown the formation of titanium sub-oxide in addition to titanium aluminides and alumina. Figure 11 shows the XRD pattern of the composite material formed from nanopowders, whereas Figure 12 shows the XRD pattern of the composite synthesized from micron aluminum and nano-size titanium dioxide reactants.

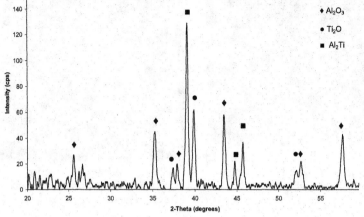

Figure 11. X-ray diffraction pattern of titanium aluminide/alumina composite synthesized from nano-size reactants.

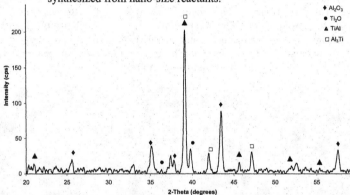

Figure 12. X-ray diffraction pattern of titanium aluminides/alumina composite synthesized from micron-size reactants.

Up to now, it is not clear why nano-size reactants lead to the formation of Al_2Ti phase and micron-size reactants resulted in the formation of mixed TiAl and $TiAl_3$. Further research is currently underway to answer those questions.

Thermodynamic analysis of the $7Al + 3TiO_2$ reacting system has confirmed the formation of titanium sub-oxides. Therefore, future research will focus on the combustion reaction with higher concentration of aluminum in the initial mixture. Based on thermodynamic predictions the higher concentration of aluminum should lead to the formation of $TiAl_3$-Al_2O_3 composite without the presence of titanium sub-oxides.

In order to increase exothermic character of the above discussed gasless combustion reaction it was decided to add nickel nanopowder as another reactant. This ternary reacting system was processed under same conditions and resulted in the formation of composites with lower residual porosity but not entirely porosity free. X-ray analysis of the product synthesized from nanoreactants indicates the presence of titanium aluminides, titanium nickelides, and alumina (see Figure 13). This analysis indicates also the presence of titanium sub-oxides, similarly to the composites formed without the addition of nickel reactant. This observation also agrees with the thermodynamic calculations.

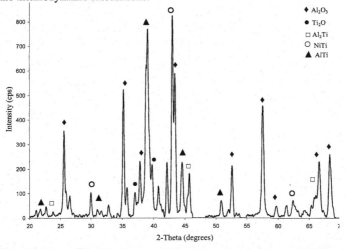

Figure 13. X-ray diffraction pattern of titanium and nickel aluminides/alumina composite synthesized from nano-size reactants.

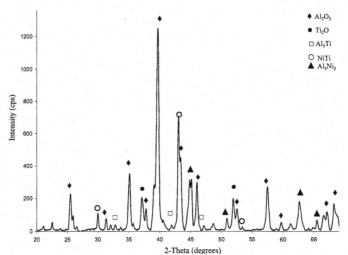

Figure 14. X-ray diffraction pattern of titanium and nickel aluminides/alumina composite synthesized from micron-size reactants.

For comparison, X-ray diffractogram of combustion synthesized composite from micron-size aluminum, and nickel, and nano-size titanium dioxide is shown in Figure 14.

CONCLUSIONS
- It was demonstrated that β-sialon powders with an average particle size in a submicron range can be synthesized by direct high pressure nitridation of silicon, and aluminum nanoreactants in the presence of alumina and a final product acting as a diluent.
- It was found that simultaneous combustion synthesis and densification technique can be successfully applied to the formation of ceramic/intermetallic composites with ultrafine microstructure when nano-size reactants are used.

ACKNOWLEDGEMENT

The authors gratefully acknowledge the financial support from the National Science Foundation. (Grant contract No. CTS-0327962).

REFERENCES
[1]G. Merzhanov, "Worldwide Evaluation and Present Status of SHS as a Branch of Modern R&D (on the 30th Anniversary of SHS)", *Int. J. SHS*, vol. 6, no. 2, pp.119-64, 1997.

[2]Z. A. Munir and U. Anselmi-Tamburini, "Self-Propagating Exothermic Reactions: The Synthesis of High-Temperature Materials by Combustion", *Mater. Sci. Rep.*, vol. 3, pp. 277-365, 1989.

[3] Alan W. Weimer (Ed.), *Carbide, Nitride and Boride Materials Synthesis and Processing*, Chapman & Hall, New York, 1997.

[4]A.G. Merzhanov, *Combustion and Plasma Synthesis of High-Temperature Materials*, Z. A. Munir and J. B. Holt, et. al. (Eds.), VCH, New York pp.1-53, 1990.

[5]A. Varma, A. S. Rogachev, A. S. Mukasyn, and S. Hwang, "Combustion Synthesis of Advanced Materials: Principles and Applications", *Adv. in Chem. Eng.*, vol. 24, pp. 79-225, 1998.

[6]J. A. Puszynski, "Kinetics and Thermodynamics of SHS Reactions", *Int. J. SHS*, vol. 10, pp. 265-293, 2001.

[7]V .L. Kvanin, V .A. Gorovoi, N .T. Balikhina, I. P. Borovinskaya, and A. G. Merzhanov, "Investigation of the Process of Forced SHS Compaction of Large-Scale Hard-Alloy Articles", *Int.J. SHS*, vol. 2, pp. 56-68, 1993.

[8]E. M. Carrillo-Heian, C. Unuvar, J. C. Gibeling, G. H. Paulino, and Z. A. Munir, "Simultaneous Synthesis and Densification of Niobium Silicide/Niobium Composites", *Scripta Mater.*, vol. 45, pp. 405-412, 2001.

[9]B. H. Rabin, G. E. Korth, and R. L. Williamson, "Fabrication of TiC-Al$_2$O$_3$ Composites by Combustion Synthesis and Subsequent Dynamic Consolidation", *Int.J.SHS*, vol. 2, pp. 336-341, 1993.

[10]A. R. Kachin and V. I. Yukhvid, "SHS for Cast Composite Materials and Pipes in the Field of Centrifugal Forces", *Int.J.SHS*, vol. 1, pp. 168-171, 1992.

[11]J.A. Puszynski and S. Miao, "Kinetic Study of Synthesis of SiC Powders and Whiskers in a Presence of KClO$_3$ or Teflon", *Int. J. SHS,* vol. 8, pp. 265-275, 1999.

[12]E. Suvaci, G. Simkovich, and G. L. Messing, "The Reaction-Bonded Aluminum Oxide Process: I, The Effect of Attrition Milling on the Solid-State Oxidation of Aluminum Powder", *J. Am. Ceram. Soc.*, vol. 83, pp. 299-305, 2000.

[13]D. E. Garcia, S. Schicker, J. Bruhn, R. Janssen, and N. Claussen, "Synthesis of Novel Niobium Aluminide-Based Composites", *J.Am.Ceram.Soc.*, vol. 80, pp. 2248-52, 1997.

[14]D. E. Garcia, S. Schicker, R. Janssen, and N. Claussen, "Nb- and Cr-Al_2O_3 Composites with Interpenetrating Networks", *J. Eur. Ceram. Soc.*, vol.18, pp. 601-605, 1998.

[15]S. Schicker, T. Erny, D .E. Garcia, R. Janssen, and N. Claussen, "Microstucture and Mechanical Properties of Al-assisted Sintered Fe/Al_2O_3 Cerments", J. Eur. Ceram. Soc., vol.19, pp. 2455-63, 1999.

[16]J.A. Puszynski, B. Liebig, S. Dargar, and J. Swiatkiewicz, "Use of Nanosize Reactants in SHS Processes", Int. J. SHS, vol.12(2), pp.107-119,2003.

[17]F. H. H. Valentin "The Mixing of Powders and Pastes:Some Basic Concepts," Chem.Eng.208 (5), pp. 99-104, 1967.

Microwave Processing

SUSCEPTOR INVESTIGATION FOR MICROWAVE HEATING APPLICATIONS

Gabrielle Gaustad and Joe Metcalfe
NYSCC at Alfred University
2 Pine Street
Alfred, New York 14802

Dr. Holly Shulman and Shawn Allan
Ceralink Inc.
200 North Main Street
Alfred, New York 14802

ABSTRACT

Many low-dielectric loss ceramics do not couple well with microwaves at low temperatures. In general, the dielectric losses increase with increasing temperature causing improved coupling. Materials which couple with microwaves at low temperatures, often called susceptors, are therefore used to improve initial heat transfer. In this study, four materials were selected for their use as possible susceptors: course grain and fine grain SiC from Research Microwave Systems, a course grinding grade SiC, and Hexalloy® from Saint-Gobain Advanced Ceramics. Susceptors were heated in a 1.3 kW, 2.45 GHz microwave unit (Research Microwave Systems) and their time to temperature as well as heating rate were compared. The susceptor with the highest heating rate and shortest time to temperature was used in varying masses to determine the effect of mass on heating rate and time to temperature. A constant mass was then used with varying refractory box free volume to determine this relationship.

INTRODUCTION

Many low-dielectric loss ceramics do not couple well with microwaves at low temperatures. In general, the dielectric losses increase with increasing temperature causing improved coupling. Materials which couple with microwaves at low temperatures, often called susceptors, are therefore used to improve initial heat transfer. This is often called hybrid microwave heating due to the dual nature: volumetric heating of the microwave and radiant heating of the susceptors.

Some researchers have experimented with a variety of susceptor materials ranging from simple alpha-silicon carbide[1] or nickel oxides[2] to complex systems such as $SiC-Al_2O_3/CaO$[3] or $Mg_xNi_{1-x}O$[2].

Lasri et al.[4] found that below the critical temperature of 800°C, the main heat-transfer mechanism of a $SiC-ZrO_2$ system was black-body radiation from the susceptor; above this temperature, microwave radiation is the primary source of temperature increase. Leiser and Clark[5] found that at low temperature (below 600°C), the type and amount of SiC as well as atmosphere were all significant factors in the time required to reach sintering temperature of an alumina cement/SiC susceptor composite. They also previously investigated[6] the effect of SiC structure on heating rate and found that beta-phase SiC achieved higher heating rates than alpha-phase SiC.

In previous work with zirconia, alumina and ZTA, it was shown that microwave energy efficiency could be increased and energy consumption decreased by increasing the load size for a given thermal package. [7] Higher heating rates also resulted in lower energy consumption. In this earlier work, the energy consumed was between 1 and 2 kW*hr/kg to heat 240 g at a rate of 50°C/min to 1500°C.

EXPERIMENTAL PROCEDURE

Experiments were completed using the Thermwave (Research Microwave Systems), a 1.3 kW, 2.45 GHz microwave unit (Figure 1). Temperature was measured with a type S thermocouple. Four materials, shown in Figure 2, were selected for their use as possible susceptors: course grain and fine grain SiC from Research Microwave Systems, a course grinding grade SiC, and Hexalloy® from Saint-Gobain Advanced Ceramics (summarized in Table I).

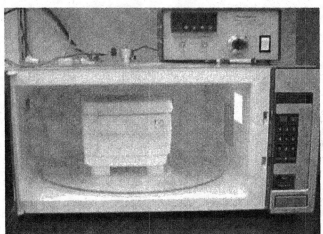

Figure 1. ThermWave (Research Microwave Systems) 2.45 GHz, 1.35 kW microwave unit with thermal package.

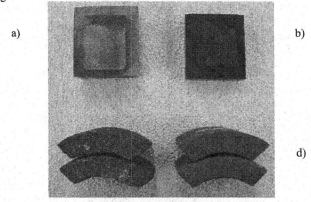

Figure 2. Susceptor Materials a) industrial grinding grade SiC, b) Hexalloy® from Saint Gobain, c) coarse grain SiC from RMS, d) fine grain SiC from RMS

Table I. Types of SiC used for susceptor performance evaluation.

SiC	Company
RMS Fine	Research Microwave Systems
RMS Coarse	Research Microwave Systems
Industrial	Coarse grained for grinding
Hexalloy®	Saint-Gobain Advanced Ceramics

For experiments completed at 50% power, the energy consumed was ~0.14 kW*hr and for runs done at 100% power the energy consumed was ~0.28 kW*hr. This was calculated using the applicator power and time per run as shown in the equation below where P is power in kilowatts, t is run time in hours, and E is energy consumed in kilowatt-hours.

$$P * t = E \qquad (1)$$

The thermal package used was a box made from Zircar fibrous alumina board with a free volume of 460 cm^3 (refer to Figure 1). Temperature was monitored with a type S thermocouple (standard deviation equal to ±1.5°C). Initial time to temperature runs were completed over an 18.5 minute duration. The susceptors were positioned in the thermal package in the center of the box, ¾" below the thermocouple. Caution was taken not to position the thermocouple too close to the samples as arcing could occur; arcing was also avoided by using a thin microwave-transparent platinum sheath over the end of the thermocouple. Data was collected at 50% (450 W) and 100% (900 W) power for all samples.

The effect of susceptor mass on heating profile and heating rate was investigated within a constant volume. Fine grain RMS SiC was used in varying masses, 40g, 80g, 130 g, and 180 g samples, in a 460 cm^3 thermal package. Each sample was run in the Thermwave at 100% (900 W) power to determine mass effects.

RMS fine grade SiC susceptor was also used for experiments with varying volume. Thermal package volumes of 368 cm^3 (small), 460 cm^3 (medium), and 606 cm^3 (large) were used with 54 grams of cold susceptor. Each run was conducted at 100% power (900 W) for 18.5 minutes.

RESULTS AND DISCUSSION
Silicon Carbide Susceptor Type
The first experiments were to compare the heating rates of four different types of SiC susceptor materials: fine grain SiC from RMS, coarse grain SiC from RMS, industrial grinding grade SiC, and Hexalloy® from Saint Gobain. For runs conducted at 50% power (~450 W), the coarse grain and fine grain SiC reached the highest temperature, 745°C and 755°C respectively, during the 18.5 minute run. The industrial grade SiC reached 650°C and the Hexalloy® reached 410°C as shown in Figure 3.
The RMS fine grain SiC kept the highest heating rates (>50°C/min.) for the longest amount of time (nearly 10 minutes) to the highest temperature (500°C) as shown in Figures 4 and 5. The coarse grain RMS performed nearly identically while the industrial grade kept high heating rates for almost 4 minutes and to 200°C. The Hexalloy® never achieved heating rates

above 50°C/min and the rate dropped rapidly within the first 2-3 minutes of the experiment. Maximum heating rates for the coarse and fine grain RMS susceptors were 72°C/minute and 70°C/minute respectively, while the grinding grade reached 59°C/minute and Hexalloy® 46°C/minute.

The comparison experiments were run again at 100% power (~900 W) and some significant differences were noted. As expected, higher temperatures and faster heating rates were achieved for all four materials seen in Figures 6-8. The fine grain and coarse grain RMS susceptors performed identically again, reaching a maximum temperature of 955°C and maximum heating rates of nearly 120°C/minute. The Hexalloy® susceptors reached a maximum temperature of 611°C and highest heating rate of 50°C showing that this material does not respond to an increase in microwave power as readily as the other types of SiC. The biggest difference was in the industrial grade SiC which began to behave closely to the fine grain and coarse grain RMS susceptors. It reached a maximum temperature of 915°C and heating rate of 110°C/minute.

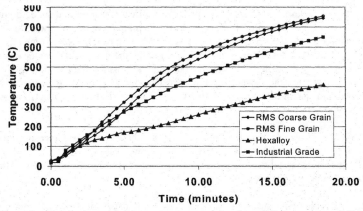

Figure 3. Comparison of time to temperature for four different types of SiC susceptor materials, constant thermal package volume, no load, at ~450 W power.

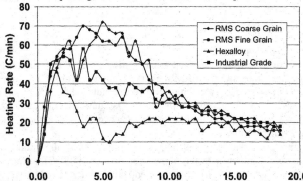

Figure 4. Comparison of heating rate versus time (minutes) for four different types of SiC susceptor materials, constant thermal package volume, no load, ~450 W power.

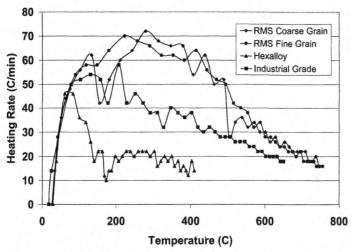

Figure 5. Comparison of heating rate versus temperature for four different types of SiC susceptor materials, constant thermal package volume, no load, ~450 W power.

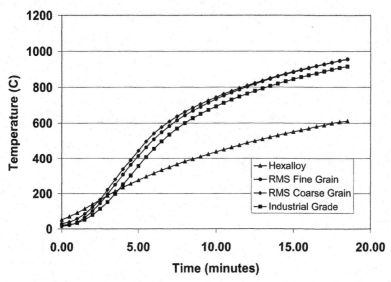

Figure 6. Time to Temperature Profile for four different types of SiC susceptor materials, constant thermal package volume, no load, ~900 W power.

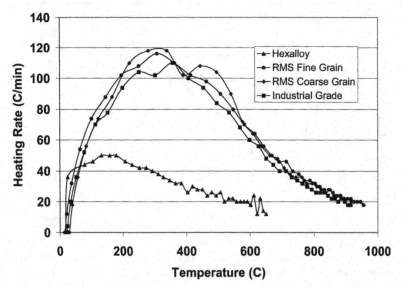

Figure 7. Heating rate versus temperature for four different types of SiC susceptor materials, constant thermal package volume, no load, ~900 W power.

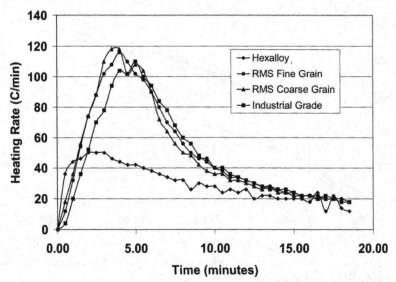

Figure 8. Heating rate versus time for four different types of SiC susceptor materials, constant thermal package volume, no load, ~900 W power.

The dielectric susceptibility of a material (χ_E) is related to its method of forming dipoles, ie. polarizability (α), the number of atoms per volume (N), the overall electric field (E), the permittivity of free space (ε_o=8.85 X 10^{-12} F/m), and the local electric field (E_i) by equation 2. For covalent solids, in this case silicon carbide which is 92% covalent[9], the only process of forming dipoles is electron cloud shifting or electronic polarizibility. The dielectric susceptibility is also intimately related to the dielectric constant of a material by equation 3 where k' is dielectric constant. Most covalent carbides have high dielectric losses[10], meaning they have difficulty following a switching electric field thereby creating atomic friction and resulting in heat.

$$\chi_E = [\, N * \alpha * E_i \,] / [\, \varepsilon_o * E \,] \qquad (2)$$

$$\chi_E = k' - 1 \quad (3)$$

The exceptional heating rates obtained for the fine grain and coarse grain RMS SiC are most probably due to the preferential heating of porous regions by high frequency electromagnetic fields[11]. The Hexalloy® samples were hot-pressed and therefore are very dense, the grinding grade SiC is also much denser than the RMS silicon carbides and therefore both do not heat as readily.

Susceptor Mass
Though both the fine grain and coarse grain RMS SiC susceptors performed nearly identically, the fine grain RMS was chosen for the experiments with varying volume. Masses of 40g, 80g, 130g, and 180 g were run in a constant volume thermal package with no load. The lowest mass had the highest initial heating rate, however, it drops off at higher temperatures. Therefore, each mass of susceptor has a range of temperature where it results in the fastest time to temperature. For the 40 gram mass this temperature is ~500°C, for 80 grams it is 500°C to 800°C, and above 800°C the 130g mass has the fastest time to temperature seen in Figure 9.

Maximum heating rate had an inverse linear relationship with susceptor mass as shown in Figure 10. The lowest mass of susceptor, 40 grams, had the highest heating rate (225°C/minute) while the highest mass of susceptor, 130 grams, had the lowest maximum heating rate of 145°C/minute. When rapid heating rates are the only variable of importance in the process however, it is best to go with low mass of susceptor.

Thermal Package Volume
RMS fine grade SiC susceptor was also used for experiments with varying volume. Thermal package volumes of 368 cm^3 (small), 460 cm^3 (medium), and 606 cm^3 (large) were used with 54 grams of cold susceptor. Figure 11 illustrates that the small volume thermal package experienced the fastest time to temperature as expected. The small volume thermal package was heated with maximum heating rates over 205°C/minute as shown in Figure 12, while the medium sized thermal package reached a maximum heating rate of 150°C/minute and the large volume, 100°C/minute.

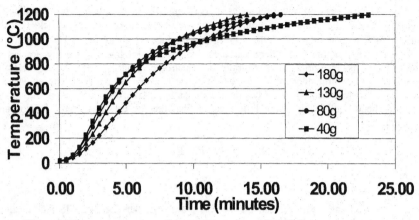

Figure 9. Time to temperature for varying masses of RMS fine grade susceptors, constant volume thermal package, no load.

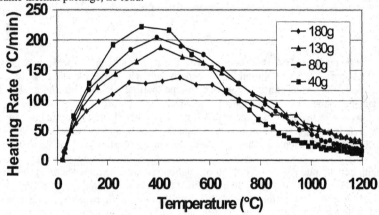

Figure 10. Heating Rate versus Temperature for varying masses of RMS fine grade susceptors, constant volume thermal package, no load.

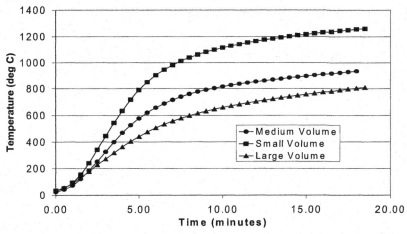

Figure 11. Temperature versus time for varying thermal package volumes, constant mass of RMS fine grade susceptor, no load.

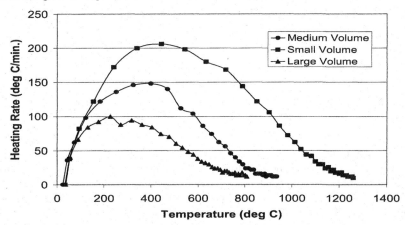

Figure 12. for varying thermal package volumes, constant mass of RMS fine grade susceptor, no load.

Energy Consumption

Due to the wide variety of reasons for using microwave heating in place of conventional, ie. time savings, finer microstructures, cost savings, etc., it is difficult to say that one susceptor type or mass is optimal for all applications. One goal that most manufacturers can relate to, and therefore is a focus of this study, is lowered energy consumption. For the experiments using varying levels of RMS fine grade susceptors (40g, 80g, 130g, and 180g), the energy consumed was calculated for runs reaching 600°C, 800°C, 1000°C, and 1200°C. During the low temperature firings (600°C and 800°C) there seems to be a linear relationship between the energy consumed and the mass of susceptors used, with the minimum mass exhibited the least amount of energy consumption (Figure 13). However, at the elevated temperatures the relationship becomes parabolic with an obvious minimum at 130g of susceptor for this size thermal package. The parabolic effect becomes even more pronounced as the temperature is elevated further.

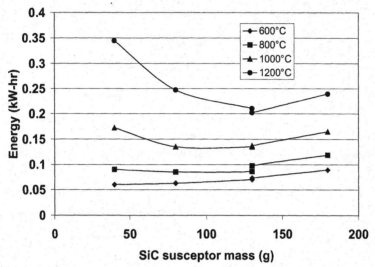

Figure 13. Energy Consumption versus RMS fine grade susceptor mass for varying temperatures, constant thermal package volume, no load.

CONCLUSIONS

Silicon carbide has many useful applications in microwave hybrid heating due to its coupling behavior at low temperatures. Of the four types of silicon carbide tested, the fine grain SiC from Research Microwave Systems had the highest heating rates. At 450 W of power, it reached rated greater than 50°C/minute and at 900 W of power, it reached heating rates of 120°C/minute. These rates were similar to the heating rates achieved by the coarse grain RMS SiC and higher than those achieved by the grinding grade SiC or the Hexalloy®. The high heating rates for the fine grain RMS SiC is probably due to microwave preferential heating of porous regions in materials.

Susceptor mass and thermal package volume are two process variables whose adjustment can be very useful in achieving maximum heating rates and minimal time to temperatures. When heating to lower temperatures (< 800°C), there appears to be a linear relationship between susceptor mass and the energy consumed reaching temperature. The lower masses of susceptor resulted in lowered energy consumption. However, at higher temperature (≥800°C), there is a parabolic relationship with a marked minimum around 130 grams of susceptor.

SUGGESTIONS FOR FUTURE WORK

There are a large variety of materials that suscept in a microwave field and therefore a broad range of materials, sizes, and shapes that could be investigated for use in hybrid microwave heating processes. The measurement of the dielectric properties (loss factor, permittivity, etc.) of the types of SiC in this study, as well as other suscepting materials of interest, would also be an excellent research topic.

REFERENCES

[1]A. Goldstein, L. Giefman, and S. Bar Ziv, "Susceptor Assisted Microwave Sintering of $MgAl_2O_4$ Powder at 2.45 GHz," *J. Mat. Sci. Lett.* **17** [12] 977-979 (1998).

[2]P. Komarenko and DE Clark, "Microwave Susceptor Materials in the $(Mg_xNi_{1-x})O$ System for High Temperature Use in Air", *Microwaves: Theory and Application in Materials Processing II.* Proc.Symp. Cincinnati, April 19-22, 1993, p.351-358. Ceram.Trans.Vol.36

[3]K.S. Leiser, R.R. Di Fiore, A.D. Cozzi, and D.E. Clark, "Microwave Heating Rates of Silicon Carbide/Alumina Cement Susceptors," *Ceram. Eng. Sci. Proc.,* **18** [4] 551-556 (1997).

[4]J. Lasri, P.D. Ramesh, and L. Schachter, "Energy Conversion During Microwave Sintering of a Multiphase Ceramic Surrounded by a Susceptor," *J. Am. Ceram. Soc.,* **83** [6] 1465-1468 (2000).

[5]K.S. Leiser and D.E. Clark, "Investigation of Microwave Behaviour of Silicon Carbide/High Alumina Cement Composites," *Ceram. Eng. Sci. Proc.* **20** [3] 103-109 (1999).

[6]K.S. Leiser and D.E.Clark, "Effects of Silicon Carbide Structure on the Microwave Properties of Silicon Carbide/Alumina Cement Suceptors," *Ceram. Eng. Sci. Proc.* **19** [4] 367-372 (1998).

[7]M.L. Fall and H. Shulman, "Comparison of Energy Consumption for Microwave Heating of Alumina, Zirconia, and Mixtures," A Presentation given at the 105th Annual American Ceramics Society National Convention in Nashville, Tennessee, April 28th, 2003.

[8]S Mandal, A Seal, S K Dalui, A K Dey, S Ghatak and A K Mukhopadhyay, Mechanical characteristics of microwave sintered silicon carbide, *Bull. Mater. Sci.,* 24 [2] 121-124 April 2001. © Indian Academy of Sciences.

[9]L. Solymar and D. Walsh, Electrical Properties of Materials, 6th ed., Oxford Science Publications, New York, 1998, p. 138.

[10]J. Batt, et. Al., "A parallel measurement programma in high temperature dielectric property measurements: an update," In *Proc. Symp. Microwave: Theory and Application In Materials Processing III, Cincinnati OH, 1-3 May 1995,* ed. D.E. Clark, D.C. Folz, S.J. Oda and R. Siberglitt. AcerS Pub, Westerville OH, 1995, pp. 313-322.

[11]A. Goldstein, W.D. Kaplan, and A. Singurindi, "Liquid assisted sintering of SiC powders by MW (2.45 GHz) heating," *J. European Ceram. Soc.* 22 1891-1896 (2002)

A COMPARISON OF THE ELECTRICAL PROPERTIES OF YTTRIA-STABILIZED ZIRCONIA PROCESSED USING CONVETIONAL, FAST-FIRE, AND MICROWAVE SINTERING TECHNIQUES

Michael Ugorek and Doreen Edwards
School of Engineering
New York State College of Ceramics
Alfred University
Alfred, NY 14802

Dr. Holly Shulman
Ceralink Inc.
Alfred, NY 14802

ABSTRACT

Microwave sintering is an energy – efficient process for sintering yttria-stabilized zirconia (YSZ). In this work, a comparative study was conducted using three different techniques – microwave, fast – fire, and traditional – to sinter 8 mol% YSZ. YSZ powders were isostatically pressed into pellets and sintered at 1350°C – 1500°C with rates ranging from 3°C/min (traditional) to 6 - 200°C/min (microwave and fast-fire) and dwell times ranging from 0 – 15min. Sintered pellets were characterized using scanning electron microscopy (SEM) to determine microstructure and impedance spectroscopy to determine high-temperature electrical properties. While the dependency of density and grain-size on sintering temperature was different for the three techniques, there were no significant differences in the relationships of density and electrical conductivity to grain size. By correlating the microstructure (grain size/ shape and the grain boundary) and high temperature electrical properties of sintered samples, a better understanding of the microwave sintering process and how this process affects final electrical properties of YSZ was determined.

INTRODUCTION

Yttria-stabilized zirconia (YSZ) is widely used as an electrolyte in solid oxide fuel cells (SOFC). Microwave sintering has been considered as an alternative to conventional sintering of YSZ, because of benefits related to lower sintering temperature, lower energy requirements, reduced thermal-mismatch stresses, and has the potential to improve the material's properties.[1-3]

In this work, three different sintering techniques – traditional (low heating rate), fast-fire, and microwave – were used to sinter ZrO_2 doped with 8 mol% Y_2O_3. The sintered samples were characterized to determine the relationships between density, grain-size, microstructure, and electrical conductivity for samples processed using the three different methods.

EXPERIMENTAL PROCEDURE

One-gram, ½-inch diameter pellets were prepared from Tosoh TZ-8YS powder by unaxial pressing at 80 MPa followed by cold isostatic pressing at 240 MPa. The pellets were sintered to 1350 - 1500°C via traditional (3°C/min), microwave (6-200°C/min), and fast-fire (6-200°C/min) techniques with dwell times ranging from 0-15min at peak temperature. Microwave sintering was conducted using a 1.3 kW, 2.45 GHz variable-power hybrid system (ThermWAVE,

Research Microwave Systems, USA). Samples were placed between two SiC susceptors inside refractory-board containers and heated using 50% power output until 500°C, followed by 100% power output to the desired sintering temperature. Figure 1 shows the heating rate attained using different refractory-board containers. Fast-fire sintering was conducted in a vertical tube furnace with computer-controlled sample feed, which allowed the heating profiles to be programmed to mimic the heating profile measured during the microwave heating runs, as shown in Figure 1. For both techniques, temperature was measured using a thermocouple placed near, but not touching, the sample. The thermocouple used during microwave heating was shielded.

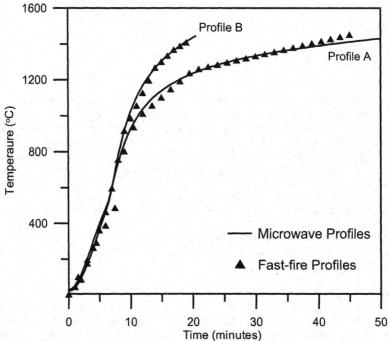

Figure 1. Experimentally measured heating profiles used during microwave and fast-fire sintering.

Pellet density was measured using Archemedes' method. SEM was conducted on the pellet surfaces and cross sections after polishing to 1 μm. Grain size was determined using a circular-intercept method.[4] Impedance spectroscopy was conducted on samples electroded with platinum paste. Impedance measurements were conducted from 100°C to 1000°C in air, using a Solatron 1260 equipped with custom control software. The samples were heated at 10°C/min, with a dwell time of 10 min at each measurement temperature. An excitation potential of 1 V was used, and frequency was scanned from 10MHz to 5Hz. The resulting spectra were analyzed using the Zview software package (Scribner and Associates, USA).

RESULTS AND DISCUSSION

The surface microstructures that were produced using all three techniques showed the same general shape, i.e. equiaxed grains in a granular aggregation. The grain sizes of the microwave and conventionally sintered samples showed a bimodal distribution whereas the grain size of the fast-fired samples was fairly uniform throughout the sample. For the microwave-sintered and conventionally sintered samples, the smaller and larger grains were segregated in localized areas, i.e. the microstructure was not uniform with respect to the two different grain sizes. These differences suggest that there was a larger amount of grain growth in the microwave-processed samples compared to the fast-fired samples, even though the microwave and fast-fired samples were heated at the same rate. Also, the microwave and traditionally processed samples produced similar microstructures even though the heating times were significantly different. Figures 2 a – 2c show typical microstructures observed on the surface of samples processed at 1450°C using the three different techniques.

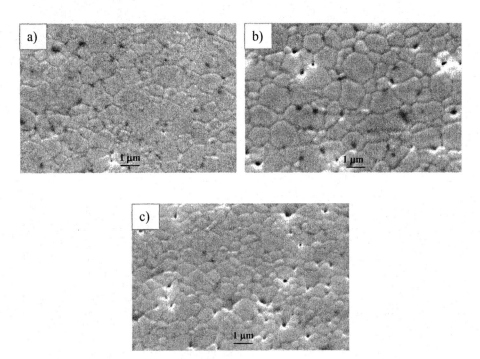

Figures 2: SEM micrographs of the surface of samples processed at 1450 °C using
a) conventional sintering, b) microwave sintering, and c) fast-fire sintering.

Figure 3a and 3b show micrographs of the cross-sectional microstructures of the microwave and fast-fired samples, respectively. Both samples were sintered at 1450°C using Heating Profile

A, as defined in Figure 1. For the microwave samples, a slight, but not statistically significant, variation in average grain size was noted over the cross sectional area; the grains at the center of the sample were slightly larger than those near the surface. While there was very little difference among the average grain sizes measured from cross sectional images, there was a significant difference between the grain size measured from the surface images compared to those measured from the cross-sectional images. For example, the a sample processed at 1450°C using the slow heating rate had average "surface" grain size of 0.7 μm compared to an average "cross-sectional" grain size of 1.0 μm. The increased grain size at the center suggests that the temperature of the interior of the microwave samples was higher than at the surface. The fast-fire samples did not show any difference in grain size when comparing the surface and the cross-sectional images. The similarity in grain size in the fast-fire samples suggests that the temperature was fairly uniform throughout the sample.

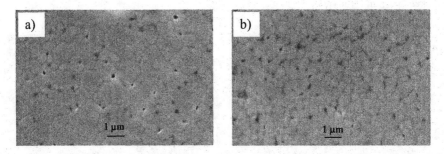

Figure 3. SEM micrographs of the cross-sections of a) a microwave sample and b) a fast-fired sample sintered at 1450 °C.

For samples processed at 1450 °C using microwave heating, there was no significant difference between the microstructures of samples processed at different heating rates. For samples fired at 1400°C, the slower heating rate (Profile A) produced a slightly larger grain size than the faster heating rate (Profile B). In all cases the porosity was located at the grain boundary regions, with little inter-granular porosity.

Figure 4 shows the density and average grain size of the YSZ samples as a function of sintering temperature. The densities of traditionally processed samples were higher than the corresponding densities of the either the microwave-processed or fast-fired samples over the entire temperature range investigated. This difference is attributed primarily to the longer sintering times associated with the slower heating rates.

The density achieved at a given sintering temperature depends strongly on the method used, especially at the lower temperatures. At 1450°C, the densities of the samples sintered using traditional sintering (low-rate) and microwave heating with Profile A are similar to each other, demonstrating that microwave sintering can achieve equivalent-density YSZ in a shorter time than can be achieved by either traditional (low-rate) sintering or fast-fire sintering.

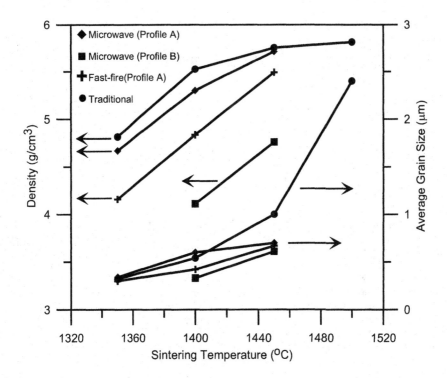

Figure 4. Density and average grain size of YSZ as a function of sintering temperature.

The density of the microwave-sintered sample is greater than that of the fast-fired sample processed with the same heating profile (Profile A). The difference is likely due to an increased average temperature for the microwave sample. Based on microstructural evidence, the internal temperature of the microwave sampled is believed to be higher than the surface temperature.

The density of the microwave sample prepared with the faster heating profile (Profile B) is significantly lower than the density of the microwave sample prepared with the slower heating profile (Profile A). Also, the average grain size of the sample microwave processed using Profile B is smaller than that of samples processed with the lower heating rate. These differences are attributed to the differences in sintering time.

At 1350°C, there is no notable difference in the grain size of the samples. With increasing sintering temperature, the traditionally fired samples shows a larger increase in grain size than do the samples processed with the other two methods. As with the differences in density, this increase in grain size can be attributed in part to the longer sintering time associated with the lower heating rate.

As summarized in Table I, the incorporation of a dwell time increased the density of the samples, as expected.

Table I: Average density for the three sintering techniques with the incorporation of a 15 minute dwell time

Technique	Temperature (°C)	Average Density (g/cm³)	
		No Dwell	15 Minute Dwell
Microwave (Profile B)	1400	4.11	5.32
	1450	4.76	5.70
Fast-Fire (Profile B)	1400	NA	5.28
	1450	NA	5.61
Traditional	1400	5.52	5.70

Figure 5 shows representative impedance data collected at three different temperatures. The spectrum collected at 300°C (Figure 5a) shows three arcs, which correspond to grain, grain-boundary, and electrode contributions. For the spectrum collected at 550°C (Figure 5b), only two arcs are present, corresponding to grain-boundary and electrode contributions. Because the frequency response of the grain interior lies outside the measurement range, the grain contribution was modeled as an offset resistance. For the spectrum collected at 900°C (Figure 5c), only one arc, corresponding to the electrode contribution, is observed. In this case, the sample resistance (grain plus grain-boundary) is taken as the offset-resistance. The sample resistance (determined by combining the contributions of the grain and grain boundary resistances) were extracted from the spectra and used to calculate the sample conductivity.

Figure 6 is a representative plot of the conductivity data for a set of YSZ samples sintered at a given temperature using the three different methods. For ionic conductors, the conductivity is expected to depend on temperature as

$$\sigma = \frac{\sigma_o}{T} e^{-Ea/kT}$$

(1)

where σ is conductivity, σ_o is a constant, T is measurement temperature in Kelvin, E_a is activation energy, and k is Boltmann's constant. All samples exhibited a linear relationship of ln (σT) vs. 1000/T, indicating that the mechanism of conduction does not change over the entire temperature range.

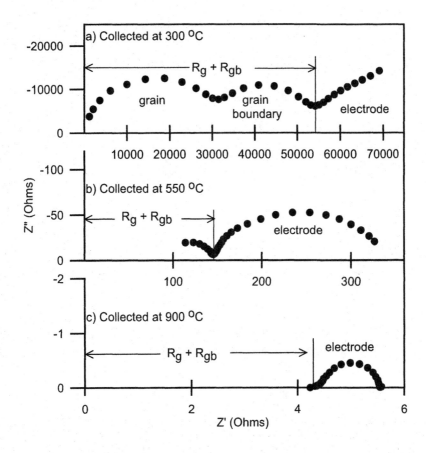

Figure 5. Representative impedance spectra collected at a) 300°C, b) 550°C, and c) 900°C.

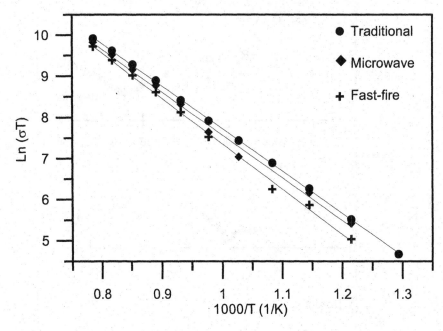

Figure 6. Electrical conductivity of YSZ as a function of measurement temperature.

Values for the activation energies and σ_o were extracted from the plots of ln (σT) vs. 1000/T and are summarized in Table II, along with the statistics of the linear least-squares fit to the data. The calculated activation energies range from 0.87 to 0.97 eV. Comparisons of the activation energies of specific samples highlight a number of trends. For a given sintering method with no dwell time, the activation energy increases with decreasing sintering temperature. The activation energies of the microwave and conventionally sintered samples were similar to each other, but slightly lower than the fast-fired samples. The activation energies of microwave samples prepared with Profile B (higher heating rate) were higher than those prepared with Profile A (lower heating rate). The activation energies of samples sintered with a dwell time are generally smaller than that of the corresponding samples sintered without a dwell time. All of these noted trends can be attributed to differences in density and grain size. In general, the activation energy for conduction decreased as sample density and grain-size increased.

Figure 7 shows the dependence of density and conductivity at 1000°C on the grain size for all three sintering techniques. All three techniques exhibit a similar density-vs.-grain-size relationships, which suggests that the relative contribution of lattice, grain-boundary, and surface diffusion in 8 mol% YSZ does not change significantly with heating rate or with the sintering method used.

Table II. Electrical Conductivity Parameters of YSZ*

Sintering Method	Sintering Temp. (°C)	σ_o (S/cm)	E_A (eV)	R^2
Traditional (no dwell)	1350	7.5×10^5	0.93	0.999
	1400	6.9×10^5	0.89	0.999
	1450	6.4×10^5	0.88	0.995
	1500	6.7×10^5	0.86	0.999
Microwave (Profile A) (no dwell)	1350	6.5×10^5	0.92	0.995
	1400	8.4×10^5	0.92	0.997
	1450	6.0×10^5	0.87	0.999
Fast-fire (Profile A) (no dwell)	1350	6.7×10^5	0.95	0.999
	1400	8.5×10^5	0.95	0.993
	1450	8.4×10^5	0.92	0.991
Microwave (Profile B) (no dwell)	1400	8.4×10^5	0.97	0.999
	1450	1.1×10^6	0.96	0.999
Traditional (15 m dwell)	1400	1.4×10^6	0.93	0.998
Microwave (Profile B) (15 m dwell)	1400	1.2×10^6	0.93	0.999
	1450	1.2×10^6	0.93	0.999
Fast Fire (Profile B) (15 m dwell)	1400	1.3×10^6	0.96	0.999

* Conductivity at any temperature can be calculated as $\sigma = (\sigma_o/T) \exp (E_A/kT)$

The conductivity of 8 mol% YSZ increases with increasing grain size, reflecting the significant contribution of grain boundaries to sample resistance. The conductivity-*vs.*-grain-size relationship is similar for all three methods, which suggests that the local electrical properties of the grain and grain-boundary regions do not depend on the sintering method used.

CONCLUSIONS
The microstructure and electrical properties of 8 mol% YSZ sintered using three different techniques were compared. Although the dependency of density and grain size on sintering temperature is different for the three techniques, the relationship of density and electrical conductivity to grain size is similar. This implies that any of the three techniques can be used to achieve 8 mol% YSZ with desired characteristics, but that the time required to achieve the desired microstructure will be different.

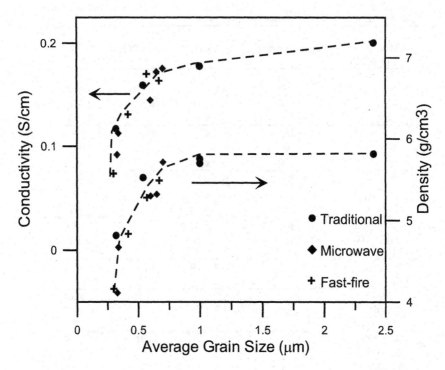

Figure 7. Electrical conductivity and density of YSZ as a function of grain size.

ACKNOWLEDGEMENTS
This work was supported, in part, by the Center for Advanced Ceramic Technology (CACT) at the New York State College of Ceramics at Alfred University.

REFERENCES
[1] S.A. Nightingale, H.K. Worner, and D.P. Dunne, "Microstructural Development during the Microwave Sintering of Yttria-Zirconia Ceramics," *Journal of the American Ceramic Society,* **80** [2] 394-400 (1997).
[2] M.A. Janney, C.L. Calhoun, and H.D. Kimrey, "Microwave Sintering of Solid Oxide Fuel Cell Materials: I, Zirconia-8 mol% Yttria," *Journal of the American Ceramic Society,* **75** [2] 341-46 (1992).
[3] F.T. Ciacchi, S.A. Nightingale, and S.P.S. Badwal, "Microwave Sintering of Zirconia-Yttria Electrolytes and Measurement of their Ionic Conductivity," *Solid State Ionics,* **86-88** 1167-72 (1996).
[4] Z. Jeffries, E. Met, A.H. Kline, and E.B. Zimmer, "The Determination of Grain Size in Metals," *Transactions,* American Institute of Mining and Metallurgical Engineers, **54** 594-607 (1917).

Sol-Gel Synthesis

BIOMORPHIC (Si, Ti, Zr) – CARBIDE SYNTHESIZED THROUGH SOL-GEL PROCESS

C. R. Rambo, J. Cao and H. Sieber

University of Erlangen-Nuremberg, Department of Materials Science, Glass and Ceramics, Martensstrasse, 5, D-91058 Erlangen, Germany.

ABSTRACT

Biomorphic carbide ceramics MeC (with Me = Si, Ti, Zr) were produced by infiltration of low viscous organo-metallic colloidal suspensions into biologically derived carbon templates and hydrolyzed in situ to form hydrogels. After hydrolysis the infiltrated templates were submitted to a high temperature pyrolysis/reaction treatment at 1600°C to promote the decomposition of the gel into metal-oxides. A carbothermal reaction of the metal-oxides with the biocarbon template finally yielded biomorphous metal carbide ceramics.

INTRODUCTION

SiC, ZrC and TiC belong to the group of high melting refractory carbides with high hardness and wear resistance that can be used for cutting tools and abrasives (1). Although ZrC is a promising candidate for a wide range of high temperature applications, during the last decade it has not been investigated like other transition metal carbides such as TiC or semi-metals like SiC (2). ZrC shows interesting physical and mechanical properties which are suitable for applications as structural components, as refractory material, in electronic devices or as substitute for SiC coating layer in the nuclear industry (2,3). Compared to SiC, the materials properties of ZrC and TiC-based ceramics are inferior, however, they are characterized by high melting point, higher hardness (TiC) and, especially a nearly metallic electrical conductivity. Tab. 1 summarizes typical properties of SiC-, ZrC- and TiC-ceramics.

Tab. 1: *Material properties of $TiC_{0.98}$, β-SiC and ZrC at room temperature (4).*

		TiC	SiC	ZrC
Melting point	[°C]	3067	2545	3250
Density	[g/cm³]	4.92	3.21	6.63
Young modulus	[GPa]	410-510	475	350-440
Hardness	[GPa]	28-35	25-28	24-27
Thermal Expansion Coefficient	[ppm/K]	7.4	3.8	7.3
Electrical resistivity at RT	[$\Omega \cdot$cm]	$18\text{-}25\ 10^{-5}$	$10^3\text{-}10^5$	$5\text{-}8\ 10^{-5}$

Among the methods that were used to produce MeC, the carbothermal reduction of MeO_2 (Eq. 1) is the most common. Other methods are solid-state reaction between C and Me (2) or mechanical activation (3).

$$MeO_2 + 3C \rightarrow MeC + 2CO \tag{1}$$

The carbothermal reduction of MeO_2 can be achieved by MeO_2/C powder mixtures at temperatures above $1200°C$ or from mixtures of carbonaceous/MeO_2-gels (5). The sol-gel method offers some advantages compared to conventional powder processing, such as lower reaction temperatures and the possibility of manufacturing porous ceramics and composites via template sol-infiltration.

The microstructures of naturally grown plants may serve as a template for the manufacturing of porous ceramics. In recent years, different biotemplating routes were developed for conversion of these structures into biomorphic ceramics (6-8). The previous work on biotemplating was mainly focused on the manufacturing of biomorphous SiC-based ceramics from natural lignocellulosic structures by different reactive processing routes: Si-melt infiltration (9), Si/SiO/ CH_3SiCl_3-vapor infiltration (10-12), SiO_2-sol infiltration (13,14) or infiltration with Si-organic monomers or polymers (15). Few reports dealt with the conversion of bioorganic materials into TiC-based ceramics, e.g. *Ashitani et al.* investigated the synthesis of TiC from wood preform materials by self-propagation high-temperature synthesis (16). *Sun et al.* produced TiC/C biomorphous ceramics by infiltration of tetrabutyl titanate into wood templates (17) and *Sieber et al.* manufactured TiC-based ceramics by Ti-vapor infiltration into carbonized pine wood performs (18).

The present work reports on the synthesis and the properties of Me-carbides (Me = Si, Zr and Ti) manufactured via infiltration of low viscosity sols into biologically derived carbon templates. The limitation of the sol-gel process for the production of the Me-carbides is discussed in terms of the intrinsic properties of the biocarbon templates and the phase forming reactions during the thermal treatment.

EXPERIMENTAL

Pine wood (*Pinus sylvestris*) was used as biological template structures for manufacturing of the highly porous, biomorphic carbide ceramics. Titanium iso-propoxide (TTIP) $Ti[OCH(CH_3)_2]_4$ (97%, *Alfa Aesar, Karlsruhe/Germany*), zirconium n-propoxide (ZNP) $Zr[O(CH_2)_2CH_3]_4$ (70%, *Alfa Aesar, Karlsruhe/Germany*) and Levasil 300 (30 vol.% SiO_2) (*Bayer, Leverkusen/Germany*) were used as sol precursors for infiltration into the biological porous templates. The pine *in natura* samples were cut in discs of approximately 1.5 cm in diameter and 0.5 cm in height, dried ($130°C$ / 2h in air) and pyrolysed at $800°C$ for 1h in N_2-atmosphere in order to decompose the polyaromatic hydrocarbon polymers of the native pine wood template into biocarbon. After pyrolysis the samples were vacuum-infiltrated with the sols. Details of the sols preparation and the infiltration process are described in previous work (19). After infiltration, the specimens were annealed at temperatures up to $1600°C$ in flowing Ar-atmosphere for 1h in order to decompose the Me-gel into MeO_2, and to promote the carbothermal reduction of the MeO_2 into MeC. The phase composition and microstructure of the biomorphous MeC carbide ceramics were analyzed by X-ray diffractometry (XRD) (*D 500, Siemens, Karlsruhe/Germany*) and scanning electron microscopy (SEM) (*Phillips XL 30*) measurements. Thermogravimetric analysis (TGA) (*Du Pont Inst., 951 Termo Analyze, Wilmington/Germany*) was applied to evaluate the weight change during the thermal processing.

RESULTS AND DISCUSSION

Fig. 1 shows the weight gain of the samples after different sol-infiltration/drying cycles. The dashed lines represent the equivalent MeO$_2$ weight gain required to achieve a stoichiometric conversion of the biocarbon template into MeC.

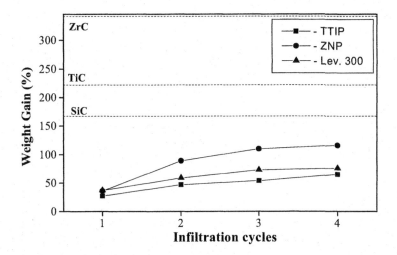

Figure 1: *Weight gain of the bio-carbon template after different infiltration cycles.*

The TGA analysis in Ar-atmosphere up to 1000°C showed that the weight loss during heating reached 20.1% for the samples infiltrated with TTIP and 19.3% for the samples infiltrated with ZNP. The major weight loss is related to the release of alcohol and water. According to Eq. 1, formation of one mole metal carbide requires three mole carbon and one mole metal oxide, which requires a theoretical weight gain of 167% for SiC, 220% for TiC and 340% for ZrC formation, respectively.

The phase composition after annealing at 1600°C in flowing Ar-atmosphere is given in Fig. 2. In the case of ZrC- and TiC-ceramics a single crystalline cubic phase was detected indicating a full conversion of the MeO$_2$ into MeC. In fact the formation of ZrC is commonly associated with an intermediary ternary oxycarbide ZrC$_x$O$_y$ formation [20]. The lattice parameter obtained for the cubic ZrC after pyrolysis was a_o = 0.4675 nm. This value is slightly smaller than the value in the JCPDS 35-784 file of pure cubic ZrC (a_o = 0.4693 nm), which indicates the presence of an oxycarbide phase ZrC$_x$O$_y$. The amount of oxygen in the oxycarbide structure decreased with increasing annealing temperature, indicating the substitution of oxygen by carbon. In the biomorphic SiC, β- and α-SiC were detected, and no peaks related to the SiO$_2$ phases (e.g. cristobalite) could be observed.

The carbothermal reduction mechanism of TiO_2 and ZrO_2 was well described by *Berger et al.* (20). It is characterized by the formation of intermediate oxides MeO_{2-x}. The main gaseous reaction products of the carbothermal reduction of TiO_2 are CO and CO_2. The carbothermal reaction mechanism for the ZrC formation was determined by *Maitre et al* [2]. The overall reaction involves two intermediary coexisting solid-gas reactions: the formation of CO and the carbothermal reduction of ZrO_2. In the case of the carbothermal reduction of silica an intermediary SiO vapor phase is formed which reacts with solid carbon.

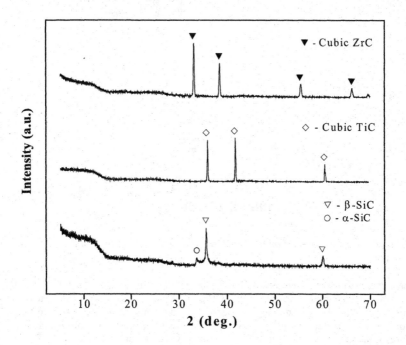

Figure 2: *XRD spectra of the biomorphic MeC ceramics after annealing at 1600°C for 1 h in flowing Ar-atmosphere.*

Despite the multiple infiltration steps, the biomorphic MeC-ceramics still contained more than 50 vol% of non-reacted carbon after infiltration and high temperature annealing. Fig. 3 shows the microstructure of the biomorphic MeC/C ceramics. SiC and TiC grain sizes are in the range of 0.5-1.5 μm and a low porosity can be found in the struts. The biomorphic ZrC-ceramic is characterized by a significantly higher fraction of unreacted carbon compared to the TiC- and SiC-ceramics.

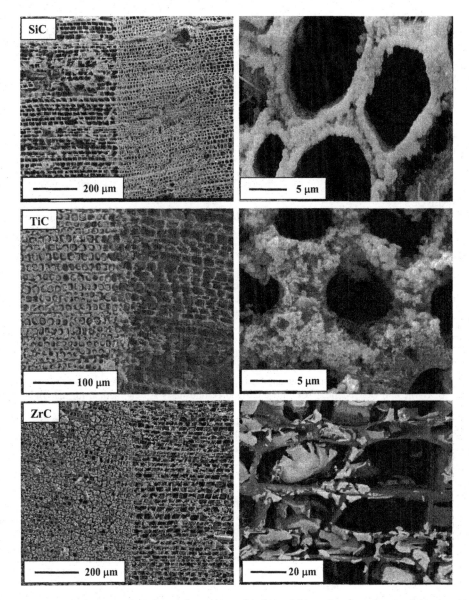

Figure 3: *SEM micrographs of the pine wood derived biocarbon template and biomorphic MeC/C ceramics.*

CONCLUSIONS

Biomorphic, highly-porous carbide ceramics, MeC with Me = Si, Ti, Zr, were produced by infiltration of low viscosity organo-metallic sols into pine wood derived biocarbon templates. Subsequent carbothermal reduction of the dried Me-oxide gels with the biocarbon template finally resulted in the formation of the biomorphic carbide phases. However, this approach was not efficient for the fully conversion of the biocarbon templates into MeC.

ACKNOWLEDGEMENTS

The authors thank CNPq-Brazil, the Volkswagen Foundation for the financial support under contract I / 73 043 and Prof. Peter Greil for helpful discussions.

REFERENCES

[1] H. Preiss, L.–M. Berger and K. Szulzewsky, "Thermal treatment of binary carbonaceous/zirconia gels and formation of Zr(C,O,N) solid solutions", *Carbon,* **34** (1) (1996) 109.

[2] A. Maitre and P. Lefort, " Solid state reaction of zirconia with carbon", *Solid State Ionics,* **104** (1997) 109.

[3] T. Tsuchida, M. Kawaguchi and K. Kodaira, " Synthesis of ZrC and ZrN in air from mechanically activated Zr-C powder mixtures", *Solid State Ionics,* **101** (1997) 149.

[4] H.O. Pierson, "Handbook of refractory carbides and nitrides", Noyes Publications, (1996).

[5] R. Koc, "Kinetics and phase evolution during carbothermal synthesis of titanium carbide from ultrafine titania/carbon mixture", *J. Mat. Sci.,* **33** (1998) 1049.

[6] A.H. Heuer, D.J. Fink, V.J. Arias, P.D. Calvert, K. Kendali, G.L. Messing, J. Blackwell, P.C. Rieke, D.H. Thompson, A.P. Wheeler, A. Veis and A.I. Caplan, "Innovative materials processing strategies: A biomimetic approach", *Science* **255** (1992) 1098.

[7] C.E. Byrne and D.E. Nagle, , "Cellulose derived composites - A new method for materials processing" *Mat. Res. Innovat.* **1** (1997) 137.

[8] P. Greil, "Biomorphous ceramics from lignocellulosics", *J. Eur. Ceram. Soc.* **21** (2001) 105.

[9] P. Greil, T. Lifka and A. Kaindl, "Biomorphic Cellular Silicon Carbide Ceramics from Wood: I. Processing and Microstructure" *J. Eur. Ceram. Soc.* <u>18</u>, (1998) 1961.

[10] E. Vogli, J. Mukerji, C. Hoffmann, R. Kladny, H. Sieber and P. Greil, "Conversion of Oak to Cellular Silicon Carbide Ceramic by Gas-Phase reaction with Silicon Monoxide", *J. Am. Cer. Soc.* **84** (2001) 1236.

[11] H. Sieber, E. Vogli, F. Müller, P. Greil, N. Popovska, H. Gerhard and G. Emig, "Gas phase processing of porous, biomorphic SiC ceramics", *Key Eng. Mat.* **206** (2002) 2013.

[12] E. Vogli, H. Sieber, and P. Greil, "Biomorphic SiC-ceramic prepared by Si-gas phase infiltration of wood", *J. Eur. Ceram. Soc.* **22** (2002) 2663.

[13] U. Vogt, A. Herzog and R. Klingner, "Porous SiC ceramics with oriented structure from natural materials", *Ceramic Engineering and Science Proceedings* **23** (4), ed. by H.-T. Lin and M. Singh, The American Ceramic Society (2002) 219.

[14] T. Ota, M. Takahashi, T. Hibi, M. Ozawa, S. Suzuki and Y. Hikichi, "Biomimetic process for producing SiC wood", *J. Am. Ceram. Soc.* **78** (1995) 3409.

[15] C. Zollfrank, R. Kladny, G. Motz, H. Sieber and P. Greil, "Manufacturing of Anisotropic Ceramics from Preceramic Polymer infiltrated Wood", *Ceramic Transactions* **129**

Innovative Processing and Synthesis of Ceramics, Glasses and Composites V, The American Ceramic Society (2002) 43.

[16]T. Ashitani, R. Tomoshige, M. Oyadomari, T. Ueno and K. Sakai, "Synthesis of titanium carbide from woody materials by self-propagating high temperature synthesis*", *Journal of the Ceramic Society of Japan* **110** (2002) 632.

[17]B. Sun, T. Fan and D. Zhang, "Porous TiC ceramics derived from wood template", *Journal of Porous Materials*, **9** (2002) 275

[18]H. Sieber, C.R. Rambo and J. Benes, "Manufacturing of biomorphous TiC-based ceramics", *Ceramic Engineering and Science Proceedings* **24** (3), ed. by W.M. Kriven and H.-T. Lin, The American Ceramic Society (2003) 135.

[19]H. Sieber, C. Rambo, J. Cao, E. Vogli and P. Greil, "Manufacturing of porous oxide ceramics by replication of wood morphologies", Key Eng. Mat. **206**, (2002) p. 2009.

[20]L.-M. Berger, W.Gruner, E. Langholf and S. Stolle, "On the mechanism of carbothermal reduction processes of TiO_2 and ZrO_2", *Int. J. Refractory Metals & Hard Materials* **17** (1999) 235.

PREPARATION OF MONODISPERSED SILICA COLLOIDS USING SOL-GEL METHOD: COSOLVENT EFFECT

Kan-Sen Chou and Chen-Chih Chen
Department of Chemical Engineering, National Tsing Hua University
101, Section 2, Kuang Fu Road, Hsinchu, Taiwan 300, Republic of China

ABSTRACT

The diameter of monodispersed silica colloids is mainly affected by the relative contribution from nucleation and growth. Once the total number of nuclei is fixed, the resultant particle size is then determined via the growth process by the total quantity of TEOS. In this work, we will demonstrate the effect of cosolvent on particle size of silica colloids, in addition to the common parameters such as NH_3 and water concentrations. Experimental results indicate that the size of silica colloids decreases with increasing polarity index of the solvents. (The definition of polarity index of cosolevent: $P'_{AB} = \phi_A P'_A + \phi_B P'_B$). Since the cosolvent is not directly involved in the preceding reactions of hydrolysis and condensation, we therefore suspect that the polarity of cosolvent would influence the solubility of intermediate $[Si(OC_2H_5)_{4-x}(OH)_x]$ and hence the supersaturation (C^*_{max}) for the nucleation process. With higher polarity of cosolvent, the number of nuclei is increased and therefore smaller silica colloids are obtained. Furthermore, the effects of ammonia and water will also be discussed by similar arguments based on relative contribution from nucleation and growth processes.

INTRODUCTION

The Stöber process on the preparation of monodispersed silica colloids by means of hydrolysis of alkyl silicates and subsequent condensation of silicic acid in alcoholic solutions using ammonia as catalyst was first published in 1968 [1]. Ever since, there have been many research groups who applied those monodispersed silica colloids as model material in various applications. Sacks and Tseng utilized those colloids to pack ordered structure membrane and investigated its sintering behavior [2,3]. Unger et al. on the other hand applied these submicron silica colloids as packing material for capillary chromatography [4,5]. In addition, there are a lot of recent investigations on using those monodispersed silica colloids to fabricate photonic crystals of 3D periodic structure [6,7]. For all these different applications, it would always be desirable to use silica particles with a specified particle size and extremely narrow distribution.

Like any other synthesis of colloids, the diameter of silica particles from the Stöber process is mainly controlled by the relative contribution from nucleation and growth. The hydrolysis and condensation reactions provide precursor species and the necessary supersaturation for the formation of particles. Once the nucleation process determines the total number of nuclei, the resultant particle size is then fixed by total quantity of TEOS [8,9]. When all the nuclei are created during the same nucleation process, we will then obtain silica colloids with narrow or even monodispersed size distributions. In general, parameters, which affect the rate of preceding reactions, would in turn affect the rate of generation of supersaturation and hence the number of nuclei formed. These parameters include reactant concentration, type of reactant, reaction temperature, concentration of catalyst (NH_3), concentration of water, as well as the choice of cosolvents.

For the sake of discussion of our experimental results, the hydrolysis and condensation reactions leading to the formation of precursor species for nucleation and growth will be briefly described as follows. During the hydrolysis reaction, the ethoxy group of TEOS reacts with the water molecule to form intermediate $[Si(OC_2H_5)_{4-x}(OH)_x]$ with hydroxyl group substituting ethoxy groups. Moreover, ammonia works as a basic catalyst to this reaction; the hydrolysis reaction is probably initiated by the attacks of hydroxyl anions on TEOS molecules [10]. The chemical reaction is expressed as follows:

$$Si(OC_2H_5)_4 + xH_2O \rightarrow Si(OC_2H_5)_{4-x}(OH)_x + xC_2H_5OH$$

Following the hydrolysis reaction, the condensation reaction occurs immediately. The hydroxyl group of intermediate $[Si(OC_2H_5)_{4-x}(OH)_x]$ reacts with either the ethoxy group of other TEOS (alcohol condensation) or the hydroxyl group of another hydrolysis intermediate (water condensation) to form Si-O-Si bridges. Furthermore, it was also claimed [11] that the rate of water condensation is thousands times faster than the alcohol condensation. Both condensation reactions can be expressed as follows:

$$\equiv Si - OC_2H_5 + HO - Si \equiv \rightarrow \equiv Si - O - Si \equiv + C_2H_5OH$$
$$\equiv Si - OH + HO - Si \equiv \rightarrow \equiv Si - O - Si \equiv + H_2O$$

The overall reaction can then be depicted as follows:

$$Si(OC_2H_5)_4 + 2H_2O \rightarrow SiO_2 + 4C_2H_5OH$$

As the reactions proceed, the intermediate (or precursor species) $[Si(OC_2H_5)_{4-x}(OH)_x]$ increases rapidly to generate supersaturation and eventually reaching a critical value of supersaturation. At this stage, the nuclei burst to form and consuming a certain extent of supersaturation [11,12]. After this point, the growth process dominates and colloids continue to grow at the expense of remaining supersaturation until all intermediates are used up. Shown in Fig. 1 is the schematic of the variation of supersaturation during the whole process.

The effect of various parameters on particle size had been reported in our previous paper [13]. The following empirical correlation was obtained.

$$D = A[H_2O]^2 \exp(-B[H_2O]^{1/2})$$
$$A = [TEOS]^{1/2}(300 + 175[NH_3] - 11.2[NH_3]^2 - 1.45[NH_3]^3)$$
$$B = 1.23 + 0.043[NH_3] - 0.0075[NH_3]^2$$

Units in the above equations are nm for D (particle diameter) and M for various concentrations. However, we felt from our past experience that the size of silica colloids is strongly influenced by the nucleation process. Therefore, in this paper, we will gather more information on the effects of catalyst (NH_3), water and cosolvent on particle size to illustrate its relation to nucleation and growth processes with the hope of precise control of particle size of silica colloids.

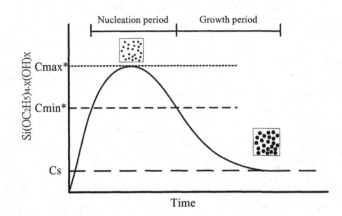

Figure 1. Schematic of nucleation and growth process for the formation of silica colloids

EXPERIMENTAL

Reagent grade chemicals of tetraethyl orthosilicate (TEOS) (98%, Acros), ammonia (28wt%, Showa), anhydrous ethanol (NASA), ethylene glycol (Showa), 1,2-dichloroethane (Tedia), methanol (Tedia) and deionized water were used as received without further purification. Here methanol, ethylene glycol and 1,2-dichloroethane serve as cosolvents. Batch process was used to prepare monodispersed silica colloids. First, solution containing appropriate quantities of anhydrous ethanol, ammonia and deionized water were stirred for 5 minutes to ensure complete mixing. Then a proper amount of TEOS was added to the above solution and the reaction proceeded at ambient temperature for 24 hours. Thereafter the colloidal solution was separated by high-speed centrifuge (Hitachi Himac CR 22G), and the silica particles were washed by deionized water for three times before characterization. The particle size of the silica particles was measured by both photon correlation spectroscopy (Nicomp 380ZLS) and scanning electron microscope (Hitachi S4700). A few samples were also characterized for their specific BET surface area (Micrometrics ASAP 2000) using nitrogen adsorption technique.

RESULTS AND DISCUSSION

First shown in Fig. 2 are some representative SEM pictures of silica colloids obtained in this work. Their sizes are clearly narrowly distributed. Relative standard deviations (i.e. standard deviation in size distribution/average diameter) are 2.04% and 7.31% respectively for these two cases. In addition, we also measured their specific surface areas using nitrogen adsorption method. Figure 3 plots the BET surface areas as a function of 1/diameter, which has the meaning of specific surface area per unit volume of particle. Also shown on this graph is the calculated external surface area as a function of 1/D for a spherical and non-porous particle. Density of silica is assumed 2.0 g/cm^3 here. Comparison of these two lines clearly indicates our silica colloids might have a rough or slightly porous surface. This finding is consistent with the TEM and SEM observation made by Costa, et al [14] that silica colloids from the Stöber process may have a thin shell with porous structure or rough surface.

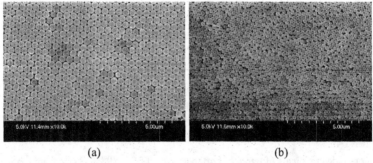

(a)   (b)

Figure 2. Representative SEM pictures of synthetic silica colloids. (a) average particle size = 498.7nm; relative standard deviation = 2.04% (experimental condition: TEOS: 0.28M; NH_3: 2M; H_2O: 14M; C_2H_5OH: 11.2M); (b) average particle size = 286.8nm; relative standard deviation = 7.31% (experimental condition: TEOS: 0.28M; NH_3: 0.37M; H_2O: 14M; C_2H_5OH: 11.9M)

Catalyst Concentration Effect (NH_3)

Figure 4 shows that the particle size increases with ammonia concentration, ranging from 0.5 to 5.0 M while TEOS and water concentrations are fixed at 0.28M and 14.0 M respectively. When ammonia is increased, both the rate of hydrolysis and condensation become faster [15]. As a result, the intermediate $[Si(OC_2H_5)_{4-x}(OH)_x]$ will be increased rapidly due to the high hydrolysis reaction; however, as it reaches the supersaturation region, the consumption rate of intermediate through condensation reaction is also relatively fast [16], which probably shortens the nucleation period. Thus, the total number of nuclei formed will be smaller, and the final particle size of synthetic silica colloids will be relatively larger as exhibited in Figure 4 under the constraint of the same total TEOS concentration in these cases.

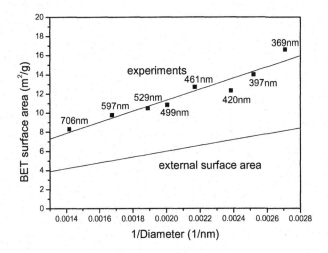

Figure 3. Correlation between BET surface area and particle size (as 1/D)

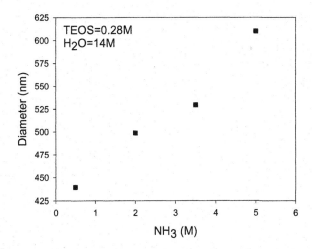

Figure 4. Effect of catalyst concentration on particle size of silica colloids

Water Concentration Effect

Next shown in Figure 5 is the effect of water concentration (1.4 M to 14 M) on particle size of the Stöber process. The role of water in this process is at least two fold. First, it is a reactant for the hydrolysis reaction. Second, it also serves as solvent, being completely miscible with ethanol. Our results show that particle size first increases with water concentration until about 7M and then decreases gradually as water concentration further increases [1,17]. This phenomenon can be explained by the two-fold role played by water. Since water is a reactant for the hydrolysis, its increase will naturally increase the rate of hydrolysis and thus a shorter period for nucleation and smaller number of nuclei formed. It will then lead to a large final particle size. However, when water concentration is further increased, the excess water can now serve a medium for the intermediate species. Due to the high polarity of water compared to ethanol, it will probably increase the solubility of the intermediate in the solution and thus prolongs the nucleation period with more nuclei produced. Accordingly, the final particle size will become smaller. At some intermediate concentration, the particle size will reach a maximum value due to the two opposing factors.

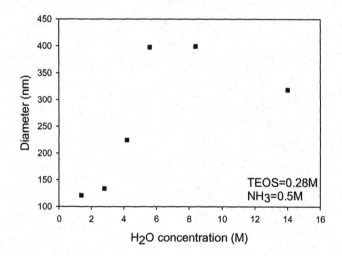

Figure 5. Effect of water concentration on particle size of silica colloids

Cosolvent Effect

Here cosolvent is another organic solvent added to ethanol to dissolve TEOS before reaction, including methanol, ethylene glycol and 1,2-dichloroethane in this work. Basically, cosolvent is not involved in the preceding reactions of hydrolysis and condensation. However, some literature [1,18] mentioned that its presence would affect the final particle size of synthetic silica colloids. Sadasivan et al. [18] used different alcohols, i.e. methanol, ethanol, propanol and butanol, as solvents to prepare silica colloids. Their results indicate that the final particle size increases with the alcohol molecule weight or decreasing polarity.

Shown in Fig. 6 are the overall results obtained in this work. Five different starting compositions were first chosen, to which different cosolvents were then added to increase or decrease the polarity of the solvent mixture and measured their final particle sizes. Clearly, our results indicated that the particle size always decreased with increasing polarity index, which is defined as follows [19]:

$$P'_{AB} = \phi_A P'_A + \phi_B P'_B$$

P'_{AB} : the polarity index of a mixture of solvents

ϕ : the volume fraction of each solvent

The polarity of the various solvents is summarized in Table 1.

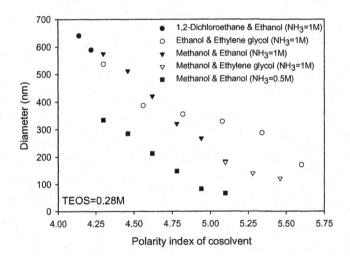

Figure 6. Effect of polarity index of cosolvent on particle size of silica colloids (TEOS = 0.28 M, H_2O = 14 M and NH_3 = 0.5 or 1.0 M)

Since the cosolvent is not considered to be involved in the chemical reactions, its effect is probably then not related with the rates of both hydrolysis and condensation. We therefore speculate that the influence of cosolvent on the formation of silica colloids is through its influence on the solubility of intermediate in the mixed solvent. The dissolution of a chemical species into a solvent depends on the molecular interaction between solvent and the chemical, which is greatly influenced by the polarity of the solvent and chemical molecules. In general, a polar species will be more soluble in a polar solvent than non-polar solvents. Besides, when the hydroxyl group gradually substitutes the ethoxy group of TEOS during hydrolysis, the intermediate reaction product becomes more polar in nature as indicated in Figure 7. It in theory also enhances its solubility in the solvent.

Table I . The polarity index of different solvent

Solvent	Polarity index
Ethylene glycol	6.9
methanol	5.1
ethanol	4.3
propanol	4.0
butanol	3.9
1,2-dichloroethane	3.5

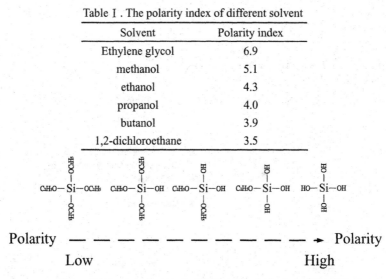

Figure 7. The polarity trend of different hydrolysis intermediates

When the solubility increases, we will then need an even higher extent of supersaturation to initiate the nucleation process. Consequently, the system may stay in the supersaturated zone for longer periods and the nucleation period will then be prolonged to produce more nuclei and hence a smaller final particle size as exhibited in Figure 6. A schematic diagram showing the effect of polarity of cosolvent on the nucleation and growth process is exhibited in Figure 8. It basically says that when more nuclei are obtained during the nucleation period, we'll then have a smaller final particle size.

Another interesting result observed from the above experiments is that the smaller particles tend to possess a wider distribution. Shown in Figure 9 are the size distributions from one such series of experiments obtained by adding different quantities of methanol to ethanol (volume ratio of methanol to ethanol being 80/20, 60/40 and 0/100 respectively) while keeping other parameters constant (TEOS = 0.28 M, NH_3 = 0.5 M, H_2O = 14.0 M). If the final smaller size is derived from a longer period of nucleation, it is then natural to expect a wider size distribution. On the contrary, in systems having a solvent with lower polarity index, the particles may experience longer period of growth. Our data suggest that growth of silica colloids from the Stöber process is self-sharpening due to the surface integration being rate-control, i.e. the distribution becomes narrower as the particles grow larger. The details will be further discussed in our future article.

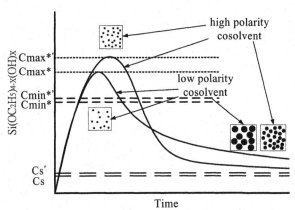

Figure 8. The schematic graph of cosolvent effect on nucleation and particle size of silica colloids

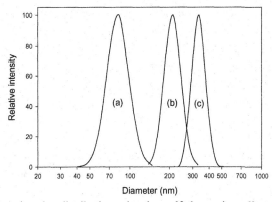

Figure 9. Representative size distributions showing self-sharpening effect of silica colloids; (a) average size 82.7 nm, relative standard deviation = 19.3%; (b) 213.0 nm; 14.8%; (c) 334.4 nm; 12.1%.

CONCLUSION

In this article, we have studied the effects of catalyst (NH$_3$), water and cosolvent on the resulting particle size of synthetic silica colloids. Our results are discussed in terms of relative contribution from nucleation and growth processes. Any parameters, which increase the rate of hydrolysis, tend to produce fewer nuclei during the nucleation process and therefore a larger particle size in the end. On the other hand, when the solvent polarity is increased, the nucleation process is probably prolonged due to higher solubility of hydrolysis intermediates to obtain more nuclei and then a smaller particle size as indicated by the effect of polarity of various cosolvents.

ACKNOWLEDGEMENT

The authors wish to thank National Science Council of Taiwan for financial support of this work (grant number NSC 92-2214-E-007-018).

REFERENCES

[1]W. Stöber, A. Fink and E. Bohn, "Controlled Growth of Monodisperse Silica Spheres in the Micron Size Range," Journal of Colloid and Interface Science, **26** 62-69 (1968).

[2]M.D. Sacks and T.Y. Tseng, "Preparation of SiO$_2$ Glass from Model Powder Compacts: I , Formation and Characterization of Powders, Suspensions, and Green Compacts," Journal of the American Ceramic Society, **67** [8] 526-532 (1984).

[3]M.D. Sacks and T.Y. Tseng, "Preparation of SiO$_2$ Glass from Model Powder Compacts: II , Sintering," Journal of the American Ceramic Society, **67** [8] 532-537 (1984).

[4]S. Ludtke, T. Adam and K.K. Unger, "Application of 0.5-μm Porous Silanized Silica Beads in Electrochromatography," Journal of Chromatography A, **786** 229-235 (1997).

[5]K.K. Unger, D. Kumar, M. Grun, G. Buchel, S. Ludtke, T. Adam, K. Schumacher and S. Renker, "Synthesis of Spherical Porous Silicas in the Micron and Submicron Size Range: Challenges and Opportunities for Miniaturized High-Resolution Chromatographic and Eletrokinetic Separations," Journal of Chromatography A, **892** 47-55 (2000).

[6]V.N. Astratov, Y.A. Vlasov, O.Z. Karimov, A.A. Kaplyanskii, Y.G. Musikhin, N.A. Bert, V.N. Bogomolov and A.V. Prokofiev, "Photonic Band Gaps in 3D Ordered fcc Silica Matrices," Physics Letters A, **222** 349-353 (1996).

[7]H. Miguez, C. Lopez, F. Meseguer, A. Blanco, L. Vazquez, R. Mayoral, M. Ocana, V. Fornes and A. Mifsud, "Photonic Crystal Properties of Packed Submicrometric SiO$_2$ Spheres," Applied Physics Letters, **71** [9] 1148-1150 (1997).

[8]T. Matsoukas and E. Gulari, "Monomer-Addition Growth with a Slow Initiation Step: A Growth Model for Silica Particles from Alkoxides," Journal of Colloid and Interface Science, **132** [1] 13-21 (1989).

[9]A.V. Blaaderen, J. V. Geest and A. Vrij, "Monodisperse Colloidal Spheres from Tetraalkoxysilande: Particle Formation and Growth Mechanism," Journal of Colloid and Interface Science, **154** [2] 481-501 (1992).

[10]C.J. Brinker and G.W. Scherer, "Sol-Gel Science: The Physics and Chemistry of Sol-Gel Processing"; pp. 97-233 Academic Press, San Diego, 1990.

[11]K.S. Kim, J.K. Kim and W.S. Kim, "Influence of Reaction Conditions on Sol-Precipitation Process Producing Silicon Oxide Particles," Ceramics International, **28** 187-194 (2002).

[12]D.L. Green, S. Jayasundara, Y.F. Lam and M.T. Harris, "Chemical Reaction Kinetics Leading to the First Stober Silica Nanoparticles-NMR and SAXS Investigation," Journal of Non-Crystalline Solids, **315** 166-179 (2003).

[13]K.S. Chou and C.C. Chen, "Preparation and Characterization of Monodispersed Silica Colloids," Advances in Technology of Materials and Materials Processing Journal, **5** [1] 31-35 (2003).

[14]C.A.L. Costa, C.A.L. Leite, E.F. de Souza and F. Galembeck, "Size Effects on the Microchemistry and Plasticity of Stöber Silica Particles: A Study Using EFTEM, FESEM, and AFM-SEPM Microscopies," Langmuir, **17** 189-194 (2001).

[15]G.H. Bogush and C.F. Zukoski IV, "Studies of the Kinetics of the Precipitation of Uniform Silica Particles through the Hydrolysis and Condensation of Silicon Alkoxides," Journal of Colloid and Interface Science, **142** [1] 1-18 (1991).

[16]S.L. Chen, P. Dong, G.H. Yang and J.J. Yang, "Kinetics of Formation of Monodisperse Colloidal Silica Particles through the Hydrolysis and Condensation of Tetraethylorthosilicate," Industrial and Engineering Chemistry Research, **35** 4487-4493 (1996).

[17]G.H. Bogush, M.A. Tracy and C.F. Zukoski IV, "Preparation of Monodisperse Silica Particles: Control of Size and Mass Fraction," Journal of Non-Crystalline Solids, **104** 95-106 (1988).

[18]S. Sadadivan, A.K. Dubey, Y. Li and D.H. Rasmussen, "Alcoholic Solvent Effect on Silica Synthesis-NMR and DLS Investigation," Journal of Sol-Gel Science and Technology, **12** 5-14 (1998).

[19]D.A. Skoog, F.J. Holler and T.A. Nieman, "Principles of Instrumental Analysis"; pp. 741-744, 5th ed. Saunders College Publishing, Florida, 1998.

Colloidal Processing

DEFINITIVE REPLACEMENT OF ORGANIC SOLVENTS BY WATER IN THE COLLOIDAL PROCESSING OF ALUMINIUM NITRIDE

José M. F. Ferreira*, Susana M. H. Olhero and Marta I.L.L. Oliveira
Department of Ceramics and Glass Engineering, University of Aveiro, CICECO,
3810-193, Aveiro, Portugal, *e-mail: jmf@cv.ua.pt

ABSTRACT

Innovative methods for protecting AlN powders against hydrolysis and for preparing stable and high concentrated ($\geq$ 50-vol.% solids) aqueous suspensions for colloidal processing (slip casting, tape casting, pressure casting, etc.) and for granulation of powders for dry pressing technologies were presented. The results achieved enabling to replace the organic-based solvents used in colloidal processing of AlN, which are much more volatile and require the control of emissions to the atmosphere, by the incombustible and non-toxicity water. Innovative features characterizing the actual method were described, which include: (i) the use of hydrophilic substances to protect the AlN surface particles against hydrolysis; (ii) the one step passivation of the hydrolysis reactions and the preparation of stable and high concentrated ($\geq$ 76-wt.%) aqueous suspensions compared with the best results already reported ($\approx$ 60-wt.%). Different consolidation methods from these suspensions could be used, namely slip casting and tape casting. Granulated powders could be produced by freeze granulation to obtain high green packing density samples by dry pressing. All the greens consolidated by the different shaping techniques could be pressureless sintered at 1750°C to achieve full dense AlN-based ceramics. The aqueous processing did not negatively affect the thermal properties of sintered bodies when compared with those processed from organic media.

INTRODUCTION

Aluminium nitride (AlN), as a substrate material in electronic packaging, has attracted considerable attention over the last two decades, due to its excellent properties.[1-3] To achieve high thermal conductivity values, the principal attribute of AlN, careful control of the processing is essential. The covalent character of the atomic bonds confers to AlN a sintering ability, inevitably forcing the use of very high sintering temperatures and the introduction of other powders in the composition, namely sintering additives, which allow sintering to proceed via a transient liquid phase by formation of a eutectic phase. Besides high temperatures, efficient sintering might also required the use of pressure in order to achieve high densities, whilst the possibility to pressureless sinter would bring clear advantages to the production of larger and more complex shaped ceramics, with a drastic decrease on production costs.

Colloidal processing methods are suited to consolidate complex shaped ceramics products and to achieve high degrees of homogeneity and particle packing that, in turn, enhance the sintering ability. In fact, colloidal shaping techniques have the capability to reduce the strength-limiting defects when comparing with dry pressing technologies.[4,5] Besides traditional processing methods, such as slip casting, tape casting, pressure casting and injection moulding, some new colloidal forming technologies have been developed in the past decade for the near-net-shape forming of complex ceramic parts, including Gel-Casting (GC), Freeze Forming (FF), Starch Consolidation (SC), Hydrolysis Assisted Solidification, (HAS), etc. The use of such performing techniques on the processing of AlN ceramics is expected to favour ceramics quality and price.[6]

Accordingly, colloidal processing is the best approach to achieve full dense materials by pressureless sintering.[7,8]

Organic based solvents are commonly used for preparation of AlN-based substrates by tape casting due to the high sensitivity of the AlN powders towards hydrolysis reactions.[9,10] However, the organic-based solvents have several disadvantages related with environmental, economic and healthy aspects. In fact, organic solvents are expensive and volatile, requiring the control of emissions to the atmosphere. Therefore, several studies have been carried out recently aiming at replacing organic solvents for water in an attempt to overcome the above referred difficulties.[11-23] Protecting AlN powders against hydrolysis reactions would enble to handle and store them in humid atmospheres, and use aqueous solutions rather than organic-based solvents to disperse AlN-based compositions. The achievement of all these goals is expected to have a tremendous positive impact at both scientific and technological levels, viabilising the aqueous colloidal processing of AlN ceramics.

"Water-resistant" AlN powders are commercially available; these materials are usually coated with hydrophobic substances, i.e., long-chain organic molecules, such as stearic acid, cetyl alcohol, n-decanoic acid, dodecylamine acid, and so on.[24-28] The problem is that the hydrophobic character of the as-treated powders makes it difficult to disperse them in water. Hydrophilic surfactants are then required to reduce the surface tension and achieve good wetting of the particles, which leads to extensive foaming of the slurry, and the need for additional anti-foaming agents.[16] All this makes the process very complex and difficult to keep under control, and as a result, air bubbles are often trapped in the suspensions and the green bodies, which means the maximum achievable solids loading is relatively low. Consequently, poor dispersion properties make difficult to obtain AlN-based ceramic materials with high green densities and microstructural homogeneity by using aqueous slip-casting methods, and explain why non-aqueous methods are still favoured for the colloidal processing of AlN components with high densities and complex shapes. Besides the hydrophobic substances mentioned above (carboxylic acids), low concentrations of some weak to poorly dissociated acids, such as *phosphoric*, H_3PO_4, or *silicic acids* in aqueous media, are known to result in a high protection efficiency of the surface powders for some days or even weeks (i.e.. for long incubation times). In the particular case of H_3PO_4, aluminium protection through anodization is known to result on impermeable and low soluble phosphate complexes, preventing the reaction. Finally, in all cases, the effectiveness of hydrolysis suppression was shown to depend on the thickness and solubility of the induced protection layer.

The present work uses complete different and innovative approaches for protecting and dispersing the AlN powders in aqueous media. Accordingly, high concentrated and stable suspensions could be prepared and used for the consolidation of complex-shaped bodies by slip casting. Adding suitable processing additives (binders and plasticizers) to the slip casting suspensions enabled the processing of AlN-based substrates by tape casting. Furthermore, granulated powders with good compacting ability for dry pressing technologies could also be prepared from the same suspensions by freeze granulation. The green bodies obtained by the different consolidation methods could be full densified by pressureless sintering at a temperature as low as 1750°C for 2 hours. The results achieved demonstrate the feasibility of a complete replacement of organic-based solvents for water in colloidal processing of AlN-based ceramics.

EXPERIMENTAL PROCEDURE
Treatments of the AlN Powder to Prevent Hydrolysis
The commercially available AlN powder (Grade C, H.C. Starck, Berlin) used in this work had an average particle size, D_{50}, $\approx$ 0.33 μm, a specific surface area of $\approx$ 6 m^2/g, and an oxygen content of 2.5-wt.% O_2, according to the information given by the supplier.
Two different approaches to protect the AlN powder against hydrolysis were conducted along this work: (i) a surface coating treatment with phosphoric acid (H_3PO_4), and (ii) a thermo-chemical connection of aluminium phosphates ($Al(H_2PO_4)_3$) to the surface of AlN particles.
The first treatment was performed by adding the AlN powder directly to a solution containing 0.2-wt.% H_3PO_4 (P) and 0.5-wt.% of an anionic surfactant (AS) (based on the solids loading of the suspension). These additives play different but complementary roles: the phosphoric acid is the protecting agent against hydrolysis reactions, while the anionic surfactant favored the dispersing of the powders.
The second surface treatment consisted of adding 100 g of AlN powder to 200 cm^3 of 2-wt.% aqueous solutions of aluminium dihydrogenphosphate, $Al(H_2PO_4)_3$, (Bindal A, TKI, Hrastnik, Slovenia). The mixtures were kept at 30, 40, 50, 60, 70 and 80°C, with continuously stirred for 15 minutes. Subsequently, the powders were allowed to settle, the supernatants were removed and the powders were dried at 70°C.

Characterization of the Powders
Hydrolysis tests were conducted using non-treated AlN powders for comparison purposes, and those treated according to the two previously described approaches. For these tests, 2 g of each AlN powder were dispersed in 100 cm^3 of water while monitoring the pH of the suspension along time during 7 hours. After that, the powders were separated from the liquid and dried at 70°C and characterized by the following techniques: (i) X-Ray diffraction (XRD) (Cu target, Kα, Philips X'Pert MPD, Netherlands) to identify the crystalline phases present in the powders before and after the hydrolysis tests; (ii) Nuclear Magnetic Ressonance (NMR) of solids was used to access the chemical bonding properties of the phosphate species to the surface of AlN powders, by recording the ^{31}P magic-angle spinning (MAS) NMR spectra on a Bruker Advance 400 Spectrometer; (iii) Electrophoresis to measure the zeta potential of all the AlN powders (non-treated, H_3PO_4-treated and $Al(H_2PO_4)_3$-treated at various temperatures), as well as of the sintering aid (CaF_2). The measurements were performed using two different instruments, a Doppler Electrophoretic light scaterring analyser (Delsa 440 SX-Coulter, UK) and a Zeta-meter (Brookhaven ZetaPlus, U.S.A), respectively. For the electrophoretic analysis, the powders were suspended in distilled water and each suspension was divided into two equal parts for increasing and decreasing pH runs (by using NaOH or HCl solutions, respectively, to adjust the pH values). Two dispersants – Dolapix CE 64 (Zschimmer & Schwarz, Germany) and Duramax 3005 (Rohm and Hass, USA) – were tested in varying amounts in order to select the most appropriate type and amount of dispersant needed to stabilize the treated AlN and the CaF_2 powders in aqueous media.

Preparation and Characterization of the Suspensions
As mentioned above, the approache to protect the AlN powder enabled to prepare a suspension with solids loading (95AlN + 5CaF$_2$ in wt.%) as high as 50-vol.% in the presence of 0.2-wt% H_3PO_4 (P) and 0.5-wt.% of the anionic surfactant (AS). The suspension was first deagglomerated by ultrasonic agitation followed by ball-milling for different times (10, 30, 60 and 120 min). Due to the new exposed surfaces originated by the deagglomerating process,

incremental amounts of 0.1-wt% of H_3PO_4 had to be added at the end of 10 and 30 min of ball milling, to guarantee the non-reactivity of the AlN powders. Rheological characterization of the suspensions was carried out with a rheometer (Carri-med, CSL500, UK) using a concentric-cylinder measuring device in the shear rate of about 0.1-550 s^{-1}.

With the AlN powder protected following the second approach, an aqueous suspension containing 50-vol.% solids (95AlN + 5CaF$_2$ in wt.%) could also be prepared after selecting the best dispersing conditions i.e., in the presence of 1-wt.% of Duramax 3005. This suspension was deagglomerated by ball milling for various time periods (7, 10 and 19 h) in a polyethylene bottle containing Teflon balls. The rheological characterization of these slips was carried out with a Rheometer (Haake, VT 500, Germany) using a MV2 system in the shear-rate range of about 4.5–450 s^{-1}.

Preparation and Characterization of the Green Slip-cast and Sintered Bodies

The packing ability of the dispersed particles during slip casting is a good indicator of the degree of dispersion achieved in the suspensions. Therefore, the required amounts of slips were poured into plastic rings that were placed on absorbent plaster plates, so as to obtain cylindrical cakes with a thickness of about 0.5 cm. The possibility of preparing AlN bodies with more complex shapes was also tested by consolidating crucibles with different sizes by slip casting. The resulting green bodies were then sintered in a graphite furnace (Degussa, Germany), using a powder bed of 80-wt.% AlN – 20-wt.% BN, in a nitrogen atmosphere at 1750°C for 2 h. The density of sintered bodies was measured according to Archimedes' method by immersion in mercury (Amsler system). The fracture surfaces of the sintered AlN bodies were analysed by Scanning Electron Microscopy (SEM) (Jeol 5800, JEOL, Japan).

RESULTS AND DISCUSSION

Characterization of the AlN powders after different pre-treatments against hydrolysis

The hydrolysis reaction kinetics of the non-treated AlN powders (NT) and the surface-treated ones with: (i) phosphoric acid (P), in absence and in the presence of an anionic surface active agent (P-AS); and (ii) surface-treated with aluminium phosphates at different temperatures, are presented in Figure 1a and 1b, respectively.

The initial pH value of the suspension made of the non-treated powder is significantly higher compared with the other ones and shows a drastic evolution with time, reaching a maximum value of about pH 10 just after 120 min. In the presence of 2-wt% of H_3PO_4 (Figure 1a), the acidity of suspension is kept below pH 4, along the whole testing period. The presence of the anionic surfactant alone delayed the hydrolyis process for more than 6 hours but revealed to be innefficient to protect AlN powder against hydrolysis reactions. The simultaneous presence of H_3PO_4 and AS provided the same degree of protection as the H_3PO_4 alone.

The evolution of pH versus time curves for the slurries of the AlN powders thermo-chemically treated at different temperatures (Figure 1b) does not show any significant pH changes over several days, indicating that hydrolysis reactions did not occur. Moreover, all the suspensions prepared with powders treated in the temperature range 50–80°C present a similar and constant pH value ($\approx$pH 6) over the whole testing period, whereas those prepared from powders treated at lower temperatures exhibit lower initial pH values, but were less constant over time. This suggests that although the treatment was effective for all the temperatures tested, lower temperatures would lead to weaker bonding of the phosphate groups to the surface of the AlN powder.

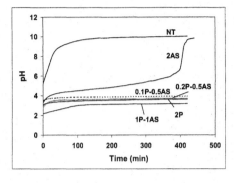 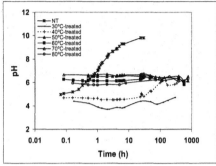

(a)   (b)

Figure 1. Evolution of pH as a function of time for the aqueous AlN suspensions: (a) in distilled water (non-treated powder –NT), or in solutions of H_3PO_4 (P) and H_3PO_4-AS (P-AS); (b) in distilled water after a thermo-chemical treatment with $Al(H_3PO_4)_3$ at various temperatures (30°C, 40°C, 50°C, 60°C, 70°C and 80°C).

The hydrolysis reactions underwent by AlN powders are known to alter their surface crystallinity and to form amorphous or crystalline aluminium hydroxide phases.[19,23] Figure 2 shows the XRD spectra obtained from the as-received powder and from those submitted to the hydrolysis tests without (NT) or with two different treatments: (i) H_3PO_4-treated (P-treated) and (ii) $Al(H_2PO_4)_3$-treated at 60°C (60°C-treated). In both treatment methods only crystalline AlN peaks were detected, like in the as-received powder, indicating the non-reactivity of the AlN powder in water. Contrarily, in the non-treated powder, bohemite and bayerite have been formed during hydrolysis. The results obtained suggest that a kind of protective coating should have been formed at the surface of AlN particles, probably constituted by some phosphate species, which inhibit hydrolysis reactions.

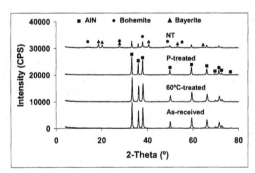

Figure 2. XRD patterns of the as-received AlN powders and of powders obtained after hydrolysis tests in different conditions: distilled water (NT); H_3PO_4-treated (P-treated); and $Al(H_2PO_4)_3$-treated at 60°C (60°C-treated).

In order to get a better understanding about the nature of the phosphate bonds at the surface of AlN powders, NMR experiments were carried out. The results displayed in Figure 3 show that the ^{31}P MAS NMR spectrum obtained from AlN H$_3$PO$_4$-treated exhibits a large peak at ca -10.7, consistent with the presence of P-O-Al environments, for example of the type P(OAl)(OH)$_3$. The large full-width-at half maximum (FWHM ca. 15ppm) of this peak may arise due to the dispersion of other types of local ^{31}P environments, for example P(OAl)$_2$(OH)$_2$ or even P(OAl)(OP)(OH)$_2$. The shorter dislocation of this large peak to more negative ppm values, the smoothness of the line spectra (less noisy) and the enhanced peak intensity observed for the thermo-chemically AlN treated with Al(H$_2$PO$_4$)$_3$, suggest the existence of stronger P-O-Al bonds.

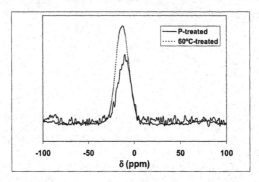

Figure 3. ^{31}P MAS NMR spectra obtained from H$_3$PO$_4$-treated (P-treated) and Al(H$_2$PO$_4$)$_3$-treated at 60°C (60°C-treated) AlN powders.

Besides the ^{31}P environments already proposed for the H$_3$PO$_4$-treated surface, other ^{31}P environments such as P(OAl)$_3$(OH) and P(OAl)$_4$ could be also envisaged for the Al(H$_2$PO$_4$)$_3$-treated powder, jointly with a higher amount of phosphates presented at the AlN surface. This results explain the higher stability of the protective coating of the AlN powder treated with Al(H$_2$PO$_4$)$_3$ when compared to the one treated with H$_3$PO$_4$.

Based on the results presented above, a schematic representation for the protective coating layers on the AlN particles, for the two different treatment methods studied, AlN-P-treated and AlN-60°C-treated, is predicted and shown on Figure 4a and 4b, respectively.

In the case of the AlN treated with phosphoric acid, different types of possible phosphate species, such us P(OAl)(OH)$_3$, P(OAl)$_2$(OH)$_2$, P(OAl)(OP)(OH)$_2$, might adsorb at the AlN particle surface togheter with the anionic surfactant through its polar head. On the other hand, in the case of the AlN powder themo-chemically treated, the temperature induced a stronger connection between the AlN surface particle and the aluminium dihidrogenophosphate species, forming a denser coating surface with interconnection between them, which present high stability against mechanical stresses during the deagglomeration/milling process.

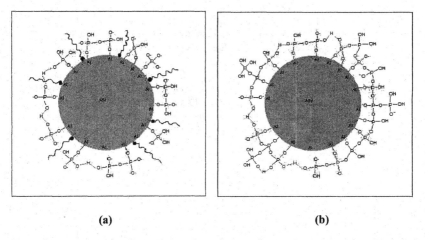

(a)   **(b)**

Figure 4. Schematic representation for the protective coating layers on the AlN particles for the two treatments studied: (a) adsorption of the phosphate acid (H_3PO_4) and anionic surfactant (AS) species and (b) thermo-chemically treated AlN powder surface with $Al(H_2PO_4)_3$.

In order to obtain information about the effects of chemisorption of H_3PO_4 (Figure 5a) and the influence of the treatment temperature (Figure 5b) on the zeta-potential of the AlN powder, on the solid liquid interface potential, measurements of zeta potential as a function of pH were performed for the as-received AlN powders, H_3PO_4-treated (P), H_3PO_4-AS-treated (P-AS) and $Al(H_2PO_4)_3$-treated at different temperatures. The results are presented in Figures 5a and 5b.

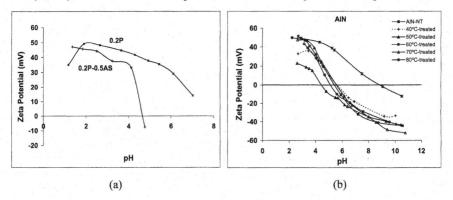

(a)   (b)

Figure 5. Zeta potential versus pH curves for as-received AlN powders and: (a) AlN treated with 0.2-wt.% H_3PO_4 (0.2P) or 0.2-wt.% H_3PO_4 + 0.5-wt.% anionic surfactant (0.2P+0.5AS); (b) AlN treated with $Al(H_3PO_4)_3$ at various temperatures (50°C, 60°C, 70°C and 80°C).

It can be seen that in absence of the surface-active agents the apparent isoelectric point (pH_{iep}) of the AlN particles is about pH 9. Such basic pH_{iep} might be due to the existence of amine groups NH and N-H$_2$, which are very common at the surface of the nitride particles. Using 0.2-wt.% of phosphoric acid solution (0.2P), as dispersing medium, caused a significant shift of the entire electrophoresis curve (Figure 5a). The shift in the pH$_{iep}$ towards acidic region means the occurrence of anionic specific adsorption, thus confirming the bonding of the phosphate ions to the surface of AlN particles. When the solution of both surface-active agents was used (0.2P + 0.5AS) as dispersing medium, the pH$_{iep}$ was further shifted towards the acidic direction being located at about pH 4.6, while zeta potential values became generally less positive, except at the lowest measured pH value. These results suggest that the carboxylic anionic surfactant has been also adsorbed at the surface of the particles and has a stronger ability to act as a surface charge modifier, compared to the phosphate ions.

After treatment in a solution of aluminium dihydrogen phosphate (Figure 5b), and irrespective of the temperature used, the pH$_{iep}$ of AlN was shifted to lower pH values: reaching a pH$_{iep}$ ≈ 4.5 after treatment at 50°C and a pH$_{iep}$ ≈ 5–5.5 for treatments at higher temperatures. The evolution of the electrophoresis curves for the powders treated at the lowest temperatures (40–50°C) is also different, showing lower zeta-potential values, both in the acidic and in the basic pH regions, compared to the other treated powders. These observations suggest that under these lower temperature conditions the phosphate groups are connected to the surface of AlN particles through weaker bonds. As a result, their partial release into the solution will increase the ionic strength of the dispersing media, therefore decreasing the zeta potential, which is consistent with the hydrolysis results. From these results we can conclude the following: for good powder protection the minimum treatment temperature should be 60°C, and after the treatment the powders should be dispersed in water at pH>8 if no surface-charge modifier agent is used to help the dispersion.

Suspensions characterization

From the electrophoretic results, the selected concentrations of H$_3$PO$_4$ and carboxylic anionic surfactant to prepare high concentrated suspensions were first set at at 0.2- and and 0.5-wt.%, respectively, based on the total mass of solids. Under these conditions, a suspension containing a solids volume fraction as high as 50-vol.% could be prepared. The starting suspension presented, however, an accentuated shear thickening behaviour as shown in Figure 6, which tend to fairly decrease along the deagglomeration time. Accordingly, a relatively fluid suspension was obtained after deagglomeration for 120 minutes, which exhibit a near Newtonian behaviour up to shear rates of about 300 s^{-1}. The shear thickening behaviour is characteristic of suspensions near the maximum packing density. Under these circumstances, the flow only occurs by increasing the average distance among particles, which in turn is counteracted by capillary forces that tend to keep particles together. This explains the increasing shear stresses required for flowing as the shear rate increases. The situation might be significantly aggravated by the presence of agglomerates, which sweep a larger volume of the suspension. The predominance of the electrostatic stabilization mechanism may also account for this behaviour at high solids loading.

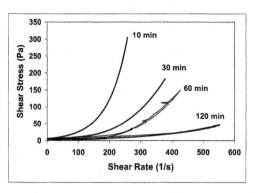

Figure 6. Flow behavior of the H_3PO_4-treated AlN aqueous suspension with 50-vol.% solids concentration after different deagglomeration times.

It was realized that the initially added amount of H_3PO_4 was insufficient to guarantee the coating of new exposed surfaces. The obtaining of deagglomerated and well stabilized suspensions required additional increments of H_3PO_4, in order to keep the coating integrity or to reform it onto the new surfaces exposed to the dispersing media as deagglomeration process progressed.[29] Accordingly, addition of two incremental amounts of 0.1-wt.% after 10 and 30 min of deagglometation. From these suspensions, AlN compacts with a relative green density as high as 71% of the theoretical density, could be obtained.

Therefore, this approach to protect the AlN powders and to prepare high concentrated suspensions might not be so reliable in terms of surface protection and may originate unpredictable and non-reproducible suspensions characteristics. Conversely, the stronger bonding of phosphate species to the surface of AlN particles achieved by the thermo-induced phosphate protection of AlN powders seems more promising and the more resistant protection layer should better outstanding the milling stresses during the deagglomeration step.

Figure 7a presents the electrophoretic curves of the thermo-chemical AlN-treated at 60°C in absence and in the presence of two different dispersants (0.6-wt.% Dolapix CE 64, or 1-wt.% Duramax 3005). Achieving high dense sintered AlN ceramics at reasonable low sintering temperatures requires the addition of sintering additives. Therefore, a CaF_2 powder was selected with this function and since all the powders in suspension must be compatible in order to avoid hetero-flocculation, the electrophoretic behaviour of CaF_2 was also studied under the same dispersion conditions. The effect of same dispersants on the zeta potential of CaF_2 particles is presented Figure 7b.

The results presented in both Figures show that Duramax 3005 is better suited to shifting the pH_{iep} of the 60°C-treated AlN and CaF_2 particles towards the acidic direction and to enhancing the negative zeta-potential values in the pH range of interest (near neutral or slightly alkaline). However, the observed shifts of the pH_{iep} towards the acidic range are much smaller for the AlN-60° treated than for the sintering additive (CaF_2). This is not surprising, since at the natural pH of the dispersant solutions the surfaces of the AlN particles are negatively charged and the driving force for the adsorption of the polyanionic groups will be lower than in the case of CaF_2.

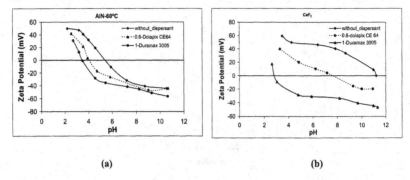

(a)   (b)

Figure 7. Zeta-potential values of the (a) AlN-60°T-treated powders and (b) CaF_2 powder, in the absence and in the presence of 0.6-wt.% of Dolapix CE 64 or 1-wt.% of Duramax 3005.

Moreover, the results of electrophoresis measurements suggest that the stabilization mechanism might be predominantly of an electrostatic nature. It is important to note that in the presence of the Duramax dispersant a good dispersion of the two components can be achieved in the pH range from 8 to 9. From these results it can be concluded that it's possible to prepare an AlN aqueous suspension in presence of the sintering additive, using Duramax 3005 as dispersant. Therefore, another improvement in using the thermo-chemical method is the possibility to work in the moderately basic pH range and use efficient commercial dispersants to stabilize the suspensions, contrarily to the previous approach that is restricted to acid pH environments.

Figure 8 shows the change with time (7, 10 and 19 hours) of the flow behaviour of aqueous suspensions prepared with 50-vol.% of the powder mixture (95AlN-60°C-treated/5CaF$_2$ in wt.%) under the selected dispersion conditions (1-wt.% Duramax 3005). All the suspensions exhibit a shear-thinning behaviour within the lower shear rate ($\dot{\gamma}$) range (up to ≈ 200 $\dot{\gamma}$ s^{-1}), followed by near-Newtonian plateaux, ending with an apparent shear thickening trend for the highest $\dot{\gamma}$ values. This behaviour is characteristic of well-dispersed and highly concentrated systems.

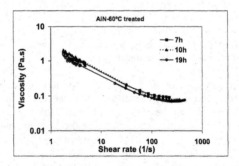

Figure 8. Flow behaviour of the aqueous suspensions containing 50-vol.% of the powders mixture (95AlN-60°C-treated/5CaF$_2$ in wt.%) after different deagglomeration times (7, 10 and 19 hours).

Once obtained highly concentrated aqueous AlN suspensions, several colloidal processing techniques could be used namely, slip casting and tape casting or to produce granulated powders by freeze granulation for dry pressing technologies. Complex shaped bodies could be consolidated by the first technique without adding further processing aids. However, the production of substrates by tape casting and the production of granules by freeze granulation required suitable additions of compatible binders and plasticizers.

All the green bodies consolidated by the different shaping methods exhibited high degrees of homogeneity and particle packing, therefore enhacing their sintering ability. Accordingly, they could be almost full densified by pressureless sintering at a temperature as low as 1750°C for 2 hours. The final properties of the processed AlN ceramics, such as the sintering density and the thermal conductivity were shown to be strongly denpendent on the processing variables, including the approaches to protect the surface of the AlN particles and the type and amount of processing additives.

Characterization of the sintered bodies

Table I shows the relative density of the sintered bodies obtained from the suspensions containing 50-vol.% solids mixed in the weight proportion of 95AlN60°C-treated/5CaF$_2$, prepared from protected AlN powders against hydrolysis following the two treatment methods described above. For comparision, the sintered density for AlN slip-cast samples obtained from an organic-based suspension that used the azeotropic mixture 60:40 of methylethylketone (MEK) – ethanol (E) as solvent and was prepared as decribed elsewhere is also reported.[30]

Table I. Sintered densities (% theoric density, TD) of the AlN samples consolidated from aqueous suspensions of AlN protected by the two surface treatment methods, and from an organic-based (MEK-E) suspension.

	Aqueous media		Organic media
	AlN-P-treated	AlN-60°C-treated	AlN-non-treated
Sintered density (%TD)	96±0.4	97±0.1	97±0.4

The density values obtained seem to reflect the rheological characteristics of the suspensions previously discussed. The shear thinning behaviour of the thermo-chemically treated AlN-based suspensions should be responsible for the higher packing density of the particles and, consequently, for the higher degree of densification achieved on sintering. It is important to note that both pre-treatments of the AlN particles' surface led to sintered density values close or equal to that obtained in the traditional organic media.

Figure 9 shows SEM micrographs at two different magnifications of a fracture surface of the 95AlN/5CaF$_2$ (in wt.%) sample sintered at 1750°C for 2 hours, revealing the high degree of densification and the well-developed polyhedral AlN crystals in the sample. Attention should be paid to the fact that the this result occurred at a sintering temperature of about 100-150°C lower than the normal sintering temperatures used for these compositions.[31-33]

Figure 9. SEM micrographs of a fracture surface of a 95AlN60°C-treated / 5CaF2 slip-cast sample after sintering at 1750°C for 2 hours.

These preliminary sintering tests were performed to demonstrate the feasibility of processing AlN ceramics with aqueous methods. Further sintering studies are necessary in order to assess the effects of various sintering aids and sintering variables (temperature, time, and atmosphere) on the degree of densification and on the final properties of AlN-based ceramics.

CONCLUSIONS

Efficient approaches to surface treat aluminium nitride powders against hydrolysis were successfully applied in the colloidal processing of AlN-based ceramics in aqueous media, using different shaping techniques. The two proposed alternative methods for protecting the surface of AlN powders enable to draw the following specific conclusions:

1. Combining suitable amounts of H_3PO_4 and AS (anionic surfactant), enables to achieve in one single step a good protection level of the surface of AlN particles against hydrolysis, as well as a good dispersing behaviour of the powders. This in turn makes possible to prepare AlN-based suspensions with 50-vol.% solids (95AlN/5CaF$_2$ in wt.%) and consolidate high green density >71% TD bodies by slip casting which achieved a density >96% TD after sintering at 1750°C for 2 hours.
2. Soaking AlN powders in an aqueous solution of aluminium dihydrogenphosphate at temperatures around 60°C offers the possibility of preparing water-resistant AlN powders. A mixture containing the well protected AlN powder and the sintering additive (95AlN/5CaF$_2$ in wt.%) could be well dispersed in a 50-vol.% solids aqueous suspension by using commercial dispersants. The high degree of homogeneity of the green bodies obtained from aqueuos slip casting enable to achieve the same degree of densification obtained for green samples consolidated from organic-based suspensions after sintering at at 1750°C for 2 hours.
3. Adding suitable types and amounts of processing additives such as binders and plasticizers also enable the consolidation of substrates by tape casting and the preparation of granulated powders for dry pressing technologies. All of these characteristics enable to definitely replace organic solvents in the processing of AlN-based ceramics by the low expensive, healthier and enviromentally friend water.

Acknowledgements

The first author wishes to thank *Fundação para a Ciência e a Tecnologia* for the financial support under the grant SFRH/BD/8754/2002.

REFERENCES

[1]Iwase, N., Ueno, F., Yasumoto, T., Asai, H. and Anzai, K., "AlN substrates and packages", *Advancing Microelectronics*, **1-2** 24 (1994).
[2]Prohaska, G. W. and Miller, G. R., "Aluminium nitride: a review of the knowledge base for physical property development", *Mat. Res. Soc. Symp. Proc.* **167** 215 (1990).
[3]P. Greil, M. Kulig, D. Hotza, H. Lange and R. Tischtau, "Aluminium nitride ceramics with high thermal conductivity from gas-phase synthesised powders", *J. Eur. Ceram. Soc.*, **13 (3)** 229 (1994).
[4]Lange, F.F. "Powder Processing Science and Technology for Increased Reliability", *J.Am. Ceram. Soc.*, **72** 3-15 (1989).
[5]Chou, C.C. & Senna, M., "Correlation between Rheological Behavior of Aqueous Suspension of Al$_2$O$_3$ and Properties of Cast Bodies: Effects of Dispersant and Ultrafine Powders", *Ceram.Bull.*, **66 (7)** 1129-33 (1987).
[6]Lewis J., "Colloidal Processing of Ceramics", *J. Am. Ceram. Soc.*, **83[10]** 2341-59 (2000).
[7]Xin Xu, Sen Mei, José M. F. Ferreira, "Fabrication of alpha-sialon sheets by tape casti and pressureless sintering", *Journal of Materials Research*, **18 (6)** 1363-1367 (2003).
[8]Xin Xu, Marta, I.L.L. Oliveira, and José J. M. Ferreira, "α-Sialon Cerami Obtained by Slip Casting and Pressureless Sintering", *J.Am. Ceram. Soc.*, **86 [2]** 366-368 (2003).
[9]Streicher E., Chartier T., Boch P., "Influence of organic components on properties of tape-cast aluminum nitride substrates", *Ceramics International*, **16** 247-252 (1990).
[10]Chartier T., Streicher E., Boch P., "Preparation and characterization of tape cast aluminum nitride substrates", *J. of Europ. Ceramic Society*, **9** 231-242 (1992).
[11]M. Oliveira, S. Olhero, J. Rocha and J.M.F. Ferreira, "Controlling hydrolysis and dispersion of AlN powders in aqueous media", *J. of Colloid and Interface Science*, **261** 456-463 (2003).
[12]Olhero S.M., Novak S., Oliveira M., Krnel K., Kosmac T., Ferreira J.M.F., "A thermo-chemical surface treatment of AlN powder for the aqueous processing of AlN ceramics", *Journal Materials Research*, **19[3]** (2004).
[13]Groat, E. A. and Mroz, T. J., "Aqueous slip casting of stabilized AlN powders", *Am. Ceram. Soc. Bull.*, **73 [11]** 75 (1994).
[14]Shimizu, Y., Hatano, J., Hyodo, T. and Egashira, M., "Ion-exchange loading of yttrium acetate as a sintering aid on aluminium nitride powder via aqueous processing", *J. Am. Ceram. Soc.* **83 [11]** 2793 (2000).
[15]Shimizu, Y., Kawanabe K., Taky, Y. Takao, Y. and Egashira, M., "AlN ceramics prepared by aqueous colloidal processing", Ceramic Processing Science and Technology, Ceramic Transactions, Edited by H. Hausner, G. L. Messing and S. Hirano, American Ceramic Society, Westerville, OH, vol.51, pg. 403-407, (1995).
[16]Kosmac, T., Krnel, K. and Kos, K., "Process for the protection of AlN powder against hydrolysis", International Patent N. WO 99/12850, March 18, 1999.

[17]Uenishi, M., Hashizume, Y. and Yokote, T., "Aluminum nitride powder having improved water resistance", Unites States Patent N. 4.923.689, May 08, 1990.

[18]Koh, Y.-H., Choi, J. and Kim, H.E., "Strengthening and prevention of oxidation of aluminium nitride by formation of a silica layer on the surface", J. Am. Ceram. Soc., 83 [2] 306 (2000).

[19]Krnel, K. and Kosmac, T., "Reactivity of aluminum nitride powder in dilute inorganic acids", J. Am. Ceram. Soc., 83 [6] 1375 (2000).

[20]Shan, H.B., Zhu, Y., Zhang, Z.T., "Surface treatment and hydrolysis kinetics of organic films coated AlN powder", British Ceramic Transactions, 98 [3] 146 (1999).

[21]Krnel K., Kosmac T., "Protection of AlN powder against hydrolysis using aluminium dihydrogen phosphate", J. Europ. Ceram. Soc., 21 2075-2079 (2001).

[22]Reetz, T., Monch, B. and Saupe, M., "Aluminum nitride hydrolysis", Cfi/Ber. DKG 69 [11/12] 464 (1992).

[23]Fukumoto, S., Hookabe, T. and Tsubakino, H., "Hydrolysis behaviour of aluminium nitride in various solutions", J. Mat. Sci., 35 2743 (2000).

[24]Egashira M., Shimizu Y., Takatsuki S., "Chemical surface treatments of aluminium nitride powder suppressing its reactivity with water", Journal of Materials Science Letters, 10 994-996 (1991).

[25]Egashira M., Shimizu Y., Takao Y., Yamaguchi R., Ishikawa Y., "Effect of carboxylic acid adsorption on the hydrolysis and sintered properties of aluminium nitride powder", J. Am. Ceram. Soc., 77 [7] 1793-1798 (1994)

[26]Xiao-Jun Luo, Xin-Rui Xu, Bao-Lin Zhang, Wen-Lan Li, Han-Rui Zhuang, "Charactesistic and dispersion of a treated AlN powder in aqueous solvent", Journal of Materials Science and Engineering A 368, 126-130 (2004).

[27]Zhang Yongheng, "Effect of surfactant on depressing the hydrolysis process for aluminium nitride powder", Materials Research Bulletin, 37 2393-2400 (2002).

[28]Kosmac T., Novak S., Krnel K., in "Proceedings of the 6th International Symposium on ceramic materials and components for engines", October 19-23, Arita-Japan, 1997.

[29]Yang, Z.Q., He L.L., Ye H.Q., "The effect of ball milling on the microstructure of ceramic AlN", Materials Science and Engineering, A323 354-357 (2002).

[30]Xu, X., Oliveira, M.I.L.L. and Ferreira J. M. F., "Effect of Solvent Composition on Dispersing Ability of Reaction Sialon Suspensions", J. Coll. Interf. Sci., 259 391-397(2003).

[31]Liang Qiao, Heping Zhou, Hao Xue, Shaohong Wang, "Effect of Y_2O_3 on low temperature sintering and thermal conductivity of AlN ceramics", J. Europ. Ceram. Soc., 23 61-67 (2003).

[32]Jackson Barrett T., Anil V. Virkar, More K. L., Dinwiddie R.B., Cutler R.A., "High thermal conductivity aluminum nitride ceramics: The effect of thermodynamic, kinetic and microstrucutural factors", J. Am. Ceram. Soc., 80 [6] 1421-35 (1997).

[33]Boch P., Glandus J.C., Jarrige J., Lecompte J.P., Mexmain J., "Sintering, Oxidation and mechanical properties of hot pressed aluminium nitride", Ceramics International, 8 [1] 34-40 (1982).

Preceramic Polymer Processing

Al$_2$O$_3$-SiC NANOCOMPOSITES BY INFILTRATION OF ALUMINA MATRIX WITH A LIQUID POLYCARBOSILANE

Dušan Galusek and Ralf Riedel
Institute of Materials Science
Darmstadt University of Technology
Petersenstraße 23
D-64287 Darmstadt
Germany

ABSTRACT

A polycarbosilane-based SiC-precursor with high ceramic yield was used for the preparation of Al$_2$O$_3$-SiC nanocomposites. The pre-sintered alumina matrix was infiltrated with a liquid precursor and subsequently heat treated under controlled conditions in inert atmosphere (Ar), in order to facilitate the ceramisation of the precursor, outgassing of the pyrolysis products and densification of the composites. The volume fraction, and grain size of the SiC particles were controlled by the volume fraction of open pores in the pre-sintered alumina matrix and by the concentration of the solution of the polymer used for the infiltration. The nanocomposites with densities ranging between 93 to 96 % of the theoretical density, and with the volume fraction of SiC ranging between 3 and 8 wt. %, and with SiC particles in nanometer range distributed within a fine-grained alumina matrix, were prepared by pressureless sintering at 1750 °C and characterized by SEM and XRD. The microstructural features were discussed with respect to the applied experimental conditions.

INTRODUCTION

Addition of particles or whiskers of a second phase is by now the most successful way of improving the mechanical properties of polycrystalline alumina-based ceramic materials. An extensive literature published on the topic quotes that the addition of silicon carbide particles (SiC$_p$) or whiskers (SiC$_w$) to polycrystalline alumina improves the strength[1 - 6], fracture toughness[1, 6 - 8], wear resistance[9 - 11], and creep resistance[12 - 14] significantly, compared to monolithic polycrystalline alumina.

Significant attention has been attracted to Al$_2$O$_3$-SiC composites by the pioneering work of Niihara whose concept of nanocomposites (addition of nano-sized particles of SiC to microcrystalline alumina matrix) allowed preparation of the Al$_2$O$_3$-SiC$_p$ materials with flexural strength exceeding 1 GPa and increased fracture toughness[1]. Despite a tremendous research effort, the reason for such an improvement remains unclear. Niihara himself suggests, that the strengthening arises due to the refinement of the microstructural scale from the order of the alumina matrix grain size to the order of the SiC interparticle spacing thus reducing the critical flaw size[1]. However, the observed toughness increase is not sufficient to account for the observed strengthening. Other authors suggest different explanations including elimination of processing flaws and suppression of enhanced grain growth[15, 16], elimination of grain pull-out during surface machining and enhanced resistance to surface defect nucleation[10], crack tip bridging with a very steep R-curve[17], high level surface compressive residual stresses induced during machining[4, 5, 18], and crack healing during annealing[4, 19]. Unlike the excellent strengthening action of the SiC nanoinclusions, the toughness increase is relatively modest.

Despite the fact that the mechanical properties of SiC-containing alumina composites are superior to those of the monolithic polycrystalline alumina, their wider commercial utilisation is limited due to a virtual lack of a cheap and reliable way allowing the preparation of fully dense, defect-free ceramic bodies with complex shape and containing homogeneously distributed reinforcing (toughening) inclusions of SiC.

The traditional preparation route consists of mixing the alumina and SiC powders in a suitable aqueous or non-aqueous media, drying, green body shaping and high temperature densification. This method has several serious drawbacks. Especially with very fine-grained (submicrometre) powders, it is difficult to prevent agglomeration of submicrometre SiC particles and to ensure homogeneous mixing of SiC and Al_2O_3. Drying the suspension is another source of agglomeration that results in uneven sintering, void and crack formation during the high temperature densification. The agglomeration encountered during the drying of suspensions can be overcome by advanced drying techniques like freeze-drying and freeze granulation, or by wet shaping techniques like slip casting, tape casting and pressure filtration.

All the techniques mentioned above produce a green body with relatively high porosity and more or less homogeneously distributed reinforcing particles SiC, which in optimum case contains no large defects or agglomerates. Careful control of the powder processing conditions allows the preparation of green body where the residual porosity can be practically completely eliminated by free sintering, but the resulting alumina matrix microstructure is relatively coarse grained (mean grain size ~ 5 μm).[20] In most cases, however, the pressureless sintering is difficult, and the SiC inclusions severely inhibit sintering by grain boundary pinning. In order to achieve full density pressure has to be applied during sintering, and hot pressing or hot isostatic pressing has to be used. This in turn limits the possibilities of producing more complex shapes, precludes the production of large series of products and increases markedly the production costs.

A concentrated effort has been therefore focused on finding alternative routes of preparation of Al_2O_3-SiC composites, which allow more homogeneous distribution of ultra-fine SiC particles within the alumina matrix and at the same time circumvent hot pressing as the principal route of densification.

The most promising method appears to be the preparation of nanocomposites by the so-called "hybrid" route utilizing SiC-forming organosilicon polymers, such as polycarbosilanes.[21 - 24] This technique is usually based on coating the alumina particles with dissolved polymer, followed by drying, cross-linking, pyrolysis and densification. The method allows formation of alumina-based nanocomposites with ultrafine particles of SiC (~12 nm) located either intra-[21, 22] or inter-granularly[23] and with high mechanical strength. As a disadvantage, in all cases reported, hot pressing was necessary to eliminate the residual porosity.

Moreover, the products of thermal decomposition of polycarbosilanes in inert atmospheres like Ar or N_2 comprise, as a result of nonstoichiometric composition of the precursors with respect to formation of SiC, not only SiC, but also free carbon.[25] Free carbon is not desirable in most of the ceramic materials, as its presence impairs the mechanical properties. However, Interrante et. al.,[26] and Riedel and Gabriel[27] reported synthesis of a polymer, which under suitable thermal processing conditions yields pure SiC without the presence of free carbon after pyrolysis at 1400-1500°C. Such polymers are now commercially available.

This work studies the preparation of Al_2O_3-SiC composites with the use of a commercially available liquid polycarbosilane, which, according to the producer data, transforms by heating in inert atmosphere directly to β-SiC with high ceramic yield and without the formation of free carbon. The liquid polymer allows its infiltration into open pores of pre-sintered alumina matrix.

The volume fraction of polymer-derived SiC and the size of SiC particles can be adjusted by the size and volume of open porosity in the pre-sintered alumina matrix, and by the concentration of the polymer solution.

EXPERIMENTAL

The ultrafine and ultrapure α-alumina powder Taimicron TM DAR (Taimei Chemicals Co., Ltd., Japan) with the average particle size of 150 nm (producer's data) was pressed axially in a steel die at 50 MPa and then isostatically at 500 MPa in order to prepare pellets with the diameter of 10 mm and of 6 mm height. The alumina green bodies were then pre-sintered in air in an electrical furnace (HTM Reetz GmbH., Berlin, Germany, model LORA 1800) under the conditions ensuring sufficient mechanical strength of the pellet and at the same time maintaining the open porosity at a required level. The conditions have been determined by a set of sintering experiments supplemented by density and porosity measurements. The density was measured by Archimedes method in water and the porosity and pore size distributions were determined by the MicroMeritics 9320 Poresizer (MicroMeritics, Norcross, GA).

Liquid polycarbosilane SP-Matrix (StarFire Systems, Watervliet, NY) was used as the infiltration agent. According to the data provided by the producer the polymer transforms upon heating in inert atmosphere (Ar) directly to β-SiC with high ceramic yield (75 – 80 wt. %, depending on the temperature). The polymer-to-ceramic conversion is accompanied by limited outgassing only (mostly hydrogen) and forms virtually no free carbon. Moreover, the polymer is liquid, soluble in aprotic solvents and can be handled in ambient environment.

Behavior of the polymer in contact with α-Al_2O_3 was examined by simultaneous thermal analysis (STA, Netzsch STA 429, Netzsch-Gerätebau GmbH, Selb, Germany), coupled with mass spectrometry (Balzers MID) in the temperature range 20 - 1500 °C on a mixture of alumina powder with 10 wt. % of the polymer pre-crosslinked for 60 minutes at 400 °C. The results were compared with the data acquired from thermal analysis of the plain polymer. The temperature intervals of development of gaseous pyrolytic products of the polymer decomposition were determined from the results of STA/MS and the heating regime was designed allowing safe outgasing of infiltrated samples during pyrolysis and sintering.

The pre-sintered alumina pellets were infiltrated with the polymer, which was used either concentrated, or dissolved in an appropriate volume of water-free cyclohexane (Sigma Aldrich, Steinheim, Germany) in order to prepare composites with 3, 5, and 8 vol. % SiC. The infiltration with the polymer solutions was carried out in static Ar atmosphere in a sealed glass container in order to avoid extended exposure of the polymer to moisture and air, and to prevent evaporation of cyclohexane. The infiltration with concentrated polymer was carried out at reduced pressure (approx. 200 Pa) in order to facilitate its penetration into the alumina matrix (viscosity 80 - 150 mN.s.m^{-2} at 20 °C). The used time of infiltration was 48 h and significantly exceeded the time the polymer requires to penetrate the whole volume of the pellet. After infiltration the excess polymer was removed from the specimen surface by paper tissue and the specimens were weighed. The solvent was evaporated by evacuation of the sample for 2 h at room temperature and the sample was re-weighed. The pellets were then placed in the crucible filled with graphite powder in order to ensure reducing conditions during pyrolysis, transferred into an electrical furnace (HTM Reetz GmbH., Berlin, Germany, model LORA 1800), and pyrolysed/sintered in flowing Ar. The sintering temperatures varied between 1550 and 1750 °C, maximum dwell time was 3 h.

Pyrolyzed and sintered samples were characterized by XRD, and SEM/EDX. XRD measurements were carried out on a STOE STADIP powder diffractometer (STOE & CIE GmbH, Darmstadt, Germany) with CuK$_\alpha$ radiation and at the 2θ angle between 30 and 70°. The SEM examinations were carried out at both fracture surfaces and the polished and thermally etched (1 h at 1500 °C in Ar) cross-sections of the specimens on a Philips XL30 high-resolution scanning electron microscope equipped with EDX analyzer (Philips, Eindhoven, The Netherlands).

The total SiC content after pyrolysis and sintering step was calculated from the content of carbon, as determined by the LECO C200 carbon content analyzer (LECO Corp., St.Joseph, MI), assuming a stoichiometric conversion of the polymer with no free carbon.

RESULTS AND DISCUSSION

Characterisation of the Alumina-Polycarbosilane Composite

The thermogravimetry data of the mixture of alumina powder with 10 wt. % of the cross-linked polymer show gradual weight decrease in four subsequent steps. Approximately 0.5 % weight loss at temperatures up to 400 °C is attributed to desorption of water from the surface of alumina particles. This is then followed by approximately 2 % weight decrease, which proceeds in two steps between 400 and 480 °C, and between 480 and 750 °C. Accompanied by exothermic effect, this step is attributed to further cross-linking of the polymer, and to the gradual conversion of the polymer to ceramic. In the temperature interval between 750 and 1200 °C the TG curve levelled off and no change of mass could be seen. The last weight loss of approximately 1 % took place at temperatures higher than 1200 °C. This result was difficult to understand, as no similar effect could be seen at the TG curve of the plain polymer (Figure 1b). In this case the mass loss was observed in the temperature intervals between 80 and 420 °C (10 %), and 420 and 900 °C (12.8 %), attributed to cross-linking, and to polymer-to-ceramic transition, respectively. Only small weight decrease (0.9 %) was detected at temperature higher then 1100 °C, caused by outgassing of the residual hydrogen. Note the alumina-polymer mixture contains only 10 wt. % of the polymer, so that the corresponding weight decrease due to the loss of the residual hydrogen would account for only 0.09 %.

The results of mass spectrometry of the alumina-polymer mixture agree well with the measured TG/DTA data, and the intervals of the development of gaseous products overlap with the temperature ranges of weight loss. The results are shown in Figure 2. The mass spectrometry measurements indicate that the decomposition of the polymer in inert atmosphere (Ar), and in presence of alumina took place in the temperature interval between 320 and 800 °C, accompanied by development of hydrogen (z = 2), and hydrocarbons with z = 12-16 (C$_1$ fragments), z = 24-28 (C$_2$ fragments), and z = 36-44 (C$_3$ fragments). Small amounts of H$_2$O fragments (z = 17, 18) also evolved, as the consequence of desorption of water from the surface of alumina particles. In this region also the units with z = 28-32 were detected, attributed to formation of the SiH$_x$ fragments as a results of cleavage of the polymer backbone, and possibly the traces of carbon monoxide as the result of pyrolytic cracking of hydrocarbons, and the reaction of precipitated carbon with water vapour:

$$C + H_2O \rightarrow CO + H_2 \tag{1}$$

The presence of water vapour contributes also to partial hydrolysis of the polycarbosilane precursor, resulting in formation of SiO$_2$. At temperatures exceeding 1200 °C the units with z = 28 were detected. The infrared analysis coupled with mass spectrometry confirmed the presence of CO, originating most likely from the carbothermal reduction of silica, which is known to proceed in some cases at temperatures lower than 1300 °C:[28, 29]

$$SiO_2 + 3C \rightarrow SiC + 2CO \qquad (2)$$

This explains at the same time the mass loss at temperatures higher than 1200 °C, as detected by the TG/DTA analysis of the alumina-polymer mixture.

Based on the STA/MS results a heating regime has been optimised, allowing safe outgassing of gaseous products from infiltrated matrix, followed by sample densification. A typical heating schedule is shown in Figure 3.

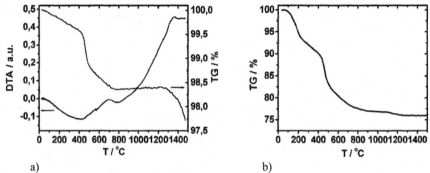

a) b)

Figure 1 DTA, TG curves of decomposition of the pre-crosslinked polymer mixed with the Al_2O_3 powder (a), and of the plain SP-Matrix polymer (b).

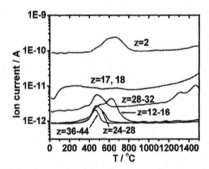

Figure 2 The temperature intervals of development of gaseous pyrolysis products as detected by mass spectrometry.

Alumina Preforms

The behaviour of the alumina green bodies during pre-sintering was monitored by the density measurements and by mercury porosimetry. The results are shown in Figures 4a, b. Figure 4a shows the decrease of the total and the open porosity of pre-sintered specimens with sintering temperature at 0 minutes dwell time (left part of the figure), and with time at 1350 °C sintering temperature (right part). Figure 4b illustrates the change of the median pore equivalent diameter. The parameters of pre-sintering of alumina preforms for infiltration experiments were then selected as follows: heating 5 °C/min to 1160 °C, with immediate cooling to room temperature at

10 °C/min. The SEM micrograph of the fracture surface of such pre-sintered alumina is shown in Figure 5. The microstructure consists of approximately equiaxed alumina grains with an average equivalent diameter of 190 nm and with significant amount of open porosity. The observed formation of necks between individual alumina grains ensures sufficient mechanical strength of the preform and prevents destruction of the body in the course of infiltration. The open porosity of the pellet was 35 vol. % and the equivalent median pore diameter was measured 52 nm. At the estimated 75 wt. % ceramic yield of the polymer, and provided that the open pores were completely filled with the undiluted polymer, this would allow preparation of the composite containing 11 vol. % SiC with the estimated equivalent median particle size SiC = 32 nm.

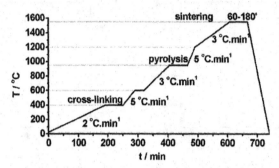

Figure 3 Typical heating schedule for pyrolysis of polymer-infiltrated alumina followed by densification step. The sintering temperature varied for different experiments.

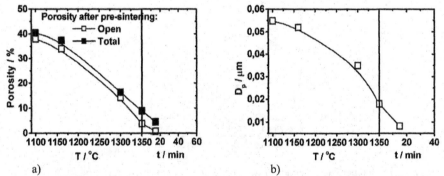

a) b)

Figure 4 The time/temperature dependence of the open and total porosity (a) and of the equivalent median size of the open pores (b) of the pre-sintered Taimicron TM-DAR.

Infiltration and characterisation

Infiltration. The amount of polymer required to prepare the composite with desired volume fraction SiC was calculated under the conditions that the ceramic yield of the polymer is 75 wt. %, and the accessible pore volume in pre-sintered alumina pellets is completely filled by the polymer, or by its cyclohexane solution. This was, however, found not to be quite true in reality. The preforms infiltrated by the viscous, concentrated polymer showed always somewhat lower weight gain than anticipated, indicating not all pore spaces were entirely filled with the polymer. On the contrary, the weight gains of pellets infiltrated by the polymer dissolved in cyclohexane showed somewhat higher weight gains than expected, most likely due to evaporation of the solvent and increase of concentration of the solution in the course of infiltration.

Pyrolysis and densification. The pyrolysis and subsequent high temperature treatment at temperatures up to 1750 °C yielded materials homogeneously dark grey or grey on the whole cross section of the pellet. In some exceptional cases (especially in pellets infiltrated by concentrated polymer) slight colour gradient could be observed, the pellets being lighter in the centre.

The carbon content measurements showed notable decrease of carbon concentrations with increased temperature of heat treatment. If stoichiometric bonding of carbon to silicon is considered, the approximate weight fraction of SiC can be calculated at each respective temperature. The results for materials infiltrated with concentrated polymer are shown in Table I. The amount of carbon decreased from equivalent of 10.1 wt. % SiC at 950 °C to equivalent of 7.3 wt. % SiC after 3 h treatment at 1650 °C, when the polymer-to-ceramic conversion is already expected to be completed.

Figure 5 SEM micrograph of the fracture surface of the alumina preform pre-sintered at 1160 °C.

Table I. Carbon content vs. the conditions of heat treatment.

Sample	Heat treatment	C, wt. %	Eq. SiC, wt. %
IP8-1	950 °C/60'	3.03 ± 0.04	10.1
IP8-2	1550 °C/60'	3.01 ± 0.08	10.0
IP8-3	1550 °C/180'	2.54 ± 0.26	8.5
IP8-4	1650 °C/180'	2.19 ± 0.13	7.3

The results of carbon content determination in specimens with various SiC contents are summarized in Table II. The data are given for specimens heat treated 3 h at 1650 °C, i.e. at the temperature where the ceramic conversion of the polymer was completed. The data are in good agreement with the weight gains measured after infiltration, indicating lower than expected contents of SiC in samples infiltrated with concentrated polymer and higher than expected SiC contents in samples infiltrated with diluted polymer.

Table II. Carbon contents of two parallel sets of samples containing 3, 5, and 8 wt. % SiC and heat treated 3 h at 1650 °C.

SiC desired,	Experiment #1		Experiment #2	
wt. %	C, wt. %	Eq. SiC, wt. %	C, wt. %	Eq. SiC, wt. %
3	1.08 ± 0.04	3.6	1.38 ± 0.01	4.6
5	1.77 ± 0.10	5.9	1.83 ± 0.02	6.1
8	2.19 ± 0.13	7.3	2.38 ± 0.05	7.9

Figure 6 shows the XRD patterns of the plain SP-Matrix polymer and the sample IP8-4, both heat treated for 3 h at 1550 °C. The plain polymer transformed directly to β-SiC and at 1550 °C the product was already crystalline. The diffraction pattern of the IP8-4 showed overlapping of the two most intensive β-SiC peaks (102 at d = 2.516 Å, 2Θ = 35.13° and 110 at d = 1.541 Å, 2Θ = 60.02°) with the peaks of α-alumina at d = 2.551 Å (2Θ = 35.61°), and 1.540 Å, respectively. However, a shoulder at d = 2.52 Å observed in the diffraction pattern of the IP8-4, and the intensity of the peak at 1.541 Å stronger than would correspond to the respective peak of α-Al$_2$O$_3$ indicate the presence of crystalline SiC in the sample.

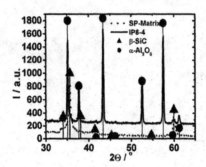

Figure 6 Diffraction patterns of the SP-Matrix polymer and the IP8-4 sample after 3 h heat treatment at 1550 °C.

The density of sintered composites as a function of sintering temperature is shown in Figure 7. All sintered samples achieved density higher than 90 % of the theoretical value after 3 h sintering at 1750 °C, the specimen with 8 wt. % SiC being 94.6 % of the theoretical density. The results are promising and indicate that complete densification of the composite by pressureless sintering can be achieved. By now, only few reports related to successful densification of Al$_2$O$_3$ - SiC composites without use of pressure, all of them quoting 5 – 7 h sintering at temperatures exceeding 1750 °C have been published.[20, 30]

Microstructure. The SEM characterisation has been carried out with both fracture surfaces and polished and etched cross sections of pyrolyzed and sintered specimens. The samples sintered 3 h at 1550 °C consist of submicrometre alumina grains covered at the surface with discrete SiC precipitates with approximate size of 50 – 100 nm (Figure 8). Practically all SiC precipitates appear to be located at boundaries between alumina grains, effectively hindering grain growth and densification by grain boundary pinning.

Longer exposure at higher sintering temperatures (1750 °C, 3 h) resulted in growth of alumina matrix grains, especially in samples with lower SiC contents (3, and 5 wt. %). The equivalent diameter of alumina grains is estimated to be $1 - 2$ μm, with occasionally observed grains up to 5 μm. These samples are also characterized by inhomogeneous distribution of nanometer-sized SiC inclusions. Figures 9a, b show two different regions of the same sample containing 5 wt. % SiC with different microstructures. For the region I a dense microstructure with homogeneously distributed and mostly intergranular SiC particles is characteristic (Figure 9a). The size of alumina grains usually does not exceed 1 μm. The region II consists of alumina grains approximately 2 μm in diameter and with grain boundaries only scarcely populated by SiC particles, which are much smaller than in region I. The region II (Figure 9b) shows the presence of pores with size comparable to that of the alumina grains and with clusters of SiC particles lining the inner walls of the pores. It may be assumed that during the evaporation of the solvent after infiltration the infiltrate tends to be drawn into larger pores where the polymer cumulates and later transforms into clusters of SiC particles. The regions depleted of SiC then sinter readily, but the pores remain preserved. It remains unclear, however, why this effect occurred in some regions, while other neighbouring regions remained unaffected, with homogeneous distribution of SiC particles.

The mixed inter-intra granular fracture mode of this sample corresponds with the observed microstructure features (Figure 9c). The fracture propagates intergranularly in less dense pore-containing regions, while in dense regions the material fails by grain cleavage.

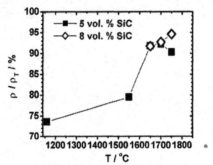

Figure 7 Dependence of density of infiltrated Al₂O₃-SiC composites on sintering temperature.

The sample with 8 vol. % precursor is, on the other hand, characteristic by homogeneous distribution of SiC particles and purely transgranular fracture (Figures 10a, b, c). The equivalent diameter of alumina grains is estimated to be ~ 1 μm. A significant fraction of intragranular SiC inclusions is located within the alumina matrix grains, which then show characteristic stepwise cleavage (Figure 10c).

Further experiments attempting the preparation of materials with higher density and the materials with high density and lower fraction of SiC particles (3 and 5 wt. %) homogeneously distributed throughout the alumina matrix are in progress.

Figure 8 Fracture surface of the Al₂O₃-(8 wt. %) SiC composite sintered for 3 h at 1550 °C.

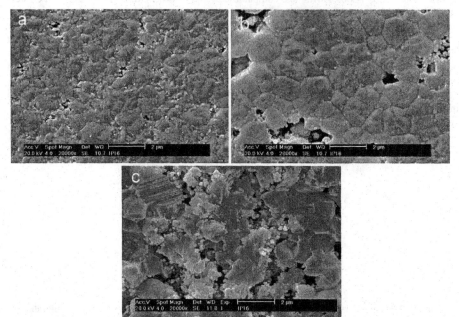

Figure 9 Polished and thermally etched cross sections of the Al₂O₃ – SiC composite sintered 3 h at 1750 °C and containing 5 wt. % SiC: region I (a), region II (b) and fracture surface of the same specimen (c).

CONCLUSIONS

Al₂O₃-SiC nanocomposites were prepared by infiltration of pre-sintered porous matrix of polycrystalline alumina with a liquid polymer. For preparation of composites containing 8 wt. % SiC the infiltrate was used as a concentrated, viscous liquid. The polymer diluted in water-free cyclohexane was used for preparation of composites containing 3 and 5 wt. % SiC. Based on TG/DTA/MS studies of decomposition of the polymer the heating regime was designed allowing safe pyrolysis and outgasing of the reaction products from alumina matrix. Three hours sintering at 1750 °C was necessary to densify the composite to 93 – 95 % of the theoretical density. The samples infiltrated with diluted polymer contained regions with lower density and with pores lined with precipitated SiC particles. The material infiltrated by the undiluted polymer was denser, with SiC precipitates distributed homogeneously throughout the alumina matrix. The precipitates located at grain boundaries effectively hindered the growth of alumina matrix grains by grain boundary pinning, and the size of alumina grains did not usually exceed 1 μm, even after 3 h sintering at 1750 °C. The material reveals transgranular fracture, with step-like cleavage of alumina grains containing intragranular inclusions of β-SiC.

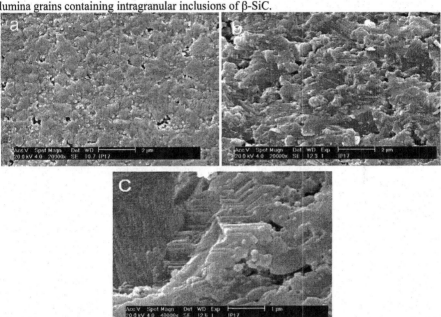

Figure 10 Polished and thermally etched cross section of the the Al_2O_3 – SiC composite sintered for 3 h at 1750 °C and containing 8 wt. % SiC (a), fracture surface of the same specimen (b) and detail of the fracture surface (c).

ACKNOWLEDGEMENT
The financial support of this work by the Alexander von Humboldt Foundation, Bonn, Germany, and by the Slovak National Grant Agency VEGA, under the contract number 2/3101/23, is gratefully acknowledged.

REFERENCES

[1] K. Niihara, "New design concept of structural ceramics – Ceramic nanocomposites", *J. Ceram. Soc. Jpn.*, **99** 974-82 (1991).

[2] R.W. Davidge, R.J. Brook, F. Cambier, M. Poorteman, A. Leriche, D. O'Sullivan, S. Hampshire and T. Kennedy, "Fabrication, properties, and modelling of engineering ceramics reinforced with nanoparticles of silicon carbide", *Br. Ceram. Trans.*, **96** 121-27 (1997).

[3] J. Perez-Rigueiro, J.Y. Pastor, J. Llorca, M. Elices, P. Miranzo and J.S. Moya, "Revisiting the mechanical behavior of alumina silicon carbide nanocomposites", *Acta Mater.*, **46** 5399-411 (1998).

[4] H.Z. Wu, C.W. Lawrence, S.G. Roberts and B. Derby, "The strength of Al_2O_3/SiC nanocomposites after grinding and annealing", *Acta Mater.*, **46** 3839-48 (1998).

[5] J. Zhao, L.C. Stearns, M.P. Harmer, H.M. Chan and G.A. Miller, "Mechanical behavior of alumina silicon-carbide nanocomposites", *J. Am. Ceram. Soc.*, **76** 503-510 (1993).

[6] D.L. Jiang and Z.R. Huang, "SiC whiskers and particles reinforced Al_2O_3 matrix composites and N_2-HIP post-treatment", *Key Eng. Mater.*, **159-60** 379-86 (1999).

[7] M.I.K. Collin and D.J. Rowcliffe, "Influence of thermal conductivity and fracture toughness on the thermal shock resistance of alumina-silicon-carbide-whisker composites", *J. Am. Ceram. Soc.*, **84** 1334-40 (2001).

[8] T. Akatsu, M. Suzuki, Y. Tanabe and E. Yasuda, "Effects of whisker content and dimensions on the R-curve behavior of an alumina matrix composite reinforced with silicon carbide whiskers", *J.Mater.Res.*, **16** 1919-27 (2001).

[9] R.W. Davidge, P.C. Twigg and F.L. Riley, "Effects of silicon carbide nano-phase on the wet erosive wear of polycrystalline alumina", *J. Eur. Ceram. Soc.*, **16** 799-802 (1996).

[10] M. Sternitzke, E. Dupas, P.C. Twigg and B. Derby, "Surface mechanical properties of alumina matrix nanocomposites", *Acta Mater.*, **45** 3963-73 (1997).

[11] H.J. Chen, W.N. Rainforth and W.E. Lee, "The wear behaviour of Al_2O_3-SiC ceramic nanocomposites", *Scripta Mat.*, **42** 555-60 (2000).

[12] A. R. De Arellano-Lopez, A. Dominguez-Rodriguez, and L. Routbort, "Microstructural constraints for creep in SiC-whisker-reinforced Al_2O_3", *Acta Mater.*, **46** 6361-73 (1998).

[13] Z.-Y. Deng, J.-L. Shi, Y.-F. Zhang, T.-R. Lai and J.-K. Guo, "Creep and creep-recovery behavior in silicon-carbide-particle-reinforced alumina", *J. Am. Ceram. Soc.*, **82** 944–52 (1999).

[14] Q. Tai and A. Mocellin, "Review: High temperature deformation of Al_2O_3-based ceramic particle or whisker composites", *Ceram. Int.*, **25** 395-408 (1999).

[15] L. Carroll, M. Sternitzke and M. Derby, "Silicon carbide particle size effects in alumina-based nanocomposites", *Acta Mater.*, **44** 4543-52 (1996).

[16] L.C. Stearns and M.P. Harmer, "Particle-inhibited grain growth in Al_2O_3-SiC: 1. Experimental results", *J. Am. Ceram. Soc.*, **79** 3013-19 (1996).

[17] T. Ohji, Y.-K. Jeong, Y.-H. Choa and K. Niihara, "Strengthening and toughening mechanisms of ceramic nanocomposites", *J. Am. Ceram. Soc.*, **81** 1453-60 (1998).

[18] I.A. Chou, H.M. Chan and M.P. Harmer, "Machining-induced surface residual stress behavior in Al_2O_3-SiC nanocomposites", *J. Am. Ceram. Soc.*, **79** 2403-409 (1996).

[19] A.M. Thompson, H.M. Chan and M.P. Harmer, "Crack healing and stress-relaxation in Al_2O_3-SiC nanocomposites", *J. Am. Ceram. Soc.*, **78** 567-71 (1995).

[20] L.C. Stearns, J. Zhao and M.P. Harmer, "Processing and microstructure development in Al_2O_3-SiC 'nanocomposites'", *J.Eur.Ceram.Soc.*, **10** 473-77 (1992).

[21] B. Su and M. Sternitzke, pp. 109-116 in *IV Euro Ceramics*, Vol. 4, Basic Science and Trends in Emerging Materials and Applications, Ed. A. Bellosi, Grupp Editoriale Faenza Editrice S.p.A., Italy, 1995.

[22] M. Sternitzke, B. Derby and R.J. Brook, "Alumina/silicon carbide nanocomposites by hybrid polymer/powder processing: Microstructures and mechanical properties", *J.Am.Ceram.Soc.*, **81** 41-48 (1998).

[23] Y. Sawai and Y. Yasutomi, "Effect of high-yield polycarbosilane addition on microstructure and mechanical properties of alumina", *J.Ceram.Soc.Jap.*, **107** 1146-50 (1999).

[24] M. Narisawa, Y. Okabe, K. Okamura and Y. Kurachi, "Synthesis of nano size dispersed silicon carbide particles by firing inorganic-organic hybrid precursors", *Key Eng. Mat.*, **159-160** 101-106 (1999).

[25] J. Bill and F. Aldinger, "Precursor-derived covalent ceramics", *Adv.Mater.*, **7** 775-87 (1995).

[26] L.V. Interrante, C.W. Whitmarsh, W. Shrewood, H.-J. Wu., R. Lewis, and G. Maciel, "Hydridopolycarbosilane precursors to silicon carbide: Synthesis, pyrolysis and application as a SiC matrix source", pp. 173-83 in NATO ASI Series, Vol. 297, *Application of Organometallic Chemistry in the Preparation and Processing of Advanced Materials*. Ed. By J.F. Harrod and R.M. Laine. Kluwe Academic Publishers, Amsterdam, The Netherlands, 1995.

[27] R. Riedel and A. Gabriel, "Synthesis of polycrystalline silicon carbide by a liquid-phase process", *Adv. Mater.*, **11** 207 (1999).

[28] V. M. Kevorkijan and A. Krizman, "Carbothermal synthesis of submicrometer β-SiC powder using double precursor reaction mixture", *Ceramic Transactions*, **51** 127-31 (1995).

[29] Z. Cheng, M. D. Sacks and C. A. Wang, "Synthesis of nanocrystalline silicon carbide powders", *Ceram. Eng. & Sci. Proc.*, **24** 23-32 (2003).

[30] H.K. Schmid, M. Aslan, S. Assmann, R. Naß and H. Schmidt, "Microstructural characterization of Al_2O_3-SiC", *J.Eur.Ceram.Soc.*, **18** 39-49 (1998).

HIGH-YIELD CERAMIC INKS FOR INK-JET PRINTING WITH PRECERAMIC POLYMERS

M. Scheffler, R. Bordia
University of Washington, Department of Materials Science & Engineering,
Box 352120, Seattle, WA 98195-2120, USA

N. Travitzky, P. Greil
University of Erlangen-Nuernberg, Department of Materials Science, Glass and Ceramics,
Martensstrasse 5, D-91058 Erlangen, Germany

ABSTRACT

Ceramic inks based on preceramic polymers were developed for inkjet printing with an office bubble jet printer. The base system consists of a water-crosslinkable silicone polymer, a latent water source for *in situ* water generation, and a tin catalyst. While the silicone polymer and the water source can be mixed in any proportion, the catalyst must be added separately to achieve crosslinking at room temperature within minutes. By pyrolysis in inert atmosphere at 1000 °C the system was shown to have a high ceramic yield exceeding 50 wt. %. The base system is compatible with alkanes, which makes it suitable for particulate filler, which can be used to control viscosity and tailor properties.

INTRODUCTION

Deposition of small amounts of functional materials has become a matter of intensive research during the last years. A promising and widely used technique for the fabrication of small parts with specific optical, electrical, chemical, biological or structural functionalities into well-defined locations is the inkjet printing technology[1]. After optimization of the basic requirements, mainly the viscosity and surface tension of the ink system, a wide field of materials systems can be processed. Examples include specific polymers into thin-film transistor circuits[2] and light-emitting polymer displays[3], biomolecules into biochips[4], 3D scaffolds as templates for biomedical applications[5], conductive gold tracks on substrates[6] or cobalt nanoparticles for catalytic growth of carbon nanotubes[7], or for combinatorial materials research[8]. Ceramic particle-loaded inks have been developed containing ZrO_2 or ZrO_2/Al_2O_3[9,10] and PZT-powders[11]. The filler amount in the dispersant liquid which is used as a transportation vehicle for the inkjet printing process, however, is limited. In reference[12] details of the inkjet printing process with respect to the flow process and the operating parameters of the printhead were modeled, and alumina suspensions with a volume fraction of up to 0.4 were used for ceramic green part manufacturing. An alternative route to increase the solid content is the use of a slurry consisting of a preceramic polymer and a ceramic powder dispersed in a solvent as demonstrated in reference[13].

Processing of preceramic polymers into ceramic products involves shaping of a polymer precursor, subsequent curing and pyrolysis at temperatures above 800°C. Due to the pronounced density differences between the polymer ($1 - 1.2$ g/cm^3) and the ceramic phases ($2 - 3$ g/cm^3) shrinkage of up to 70 vol. % may occur which gives rise to extensive porosity or cracking in the pyrolyzed ceramic residue. Manufacturing of ceramic parts from precereramic polymers, however, is facilitated when the polymer is loaded with a filler powder. *Inert filler powders* such as Al_2O_3, SiC, B_4C, Si_3N_4, etc. as well as *reactive fillers* like Ti, Cr, Mo, B, $MoSi_2$, etc., which may react with the solid and gaseous decomposition products of the polymer precursor to form carbides, oxides, etc., have been succesfully used to reduce the polymer-to-ceramic shrinkage and to improve the mechanical properties of non-oxide as well as oxide based polymer derived ceramics[14-16].

The process described in this paper is based on a liquid silicone polymer, which is crosslinkable in the presence of small amounts of water and a catalyst, and a latent water source for crosslinking at room temperature. The water source is initiated to form water when a catalyst is added during the inkjet printing process. The ink system was characterized with respect to the crosslinking time and reactions during processing and the ceramic yield of this basic as well as of a filler and silicone resin containing slurry.

EXPERIMENTAL PROCEDURE

A methoxymethyl(polysiloxane), also known as siliconeether (MSE-100, Wacker Silicone AG, Muenchen, Germany) and a hydroxy-terminated linear dimethylpolysiloxane (DMS-S12, Gelest Inc. Morrisville, PA, USA) were used in this study. Both liquid components were mixed in a test tube with a weight fraction of the MSE-100 $M_{MSE}=(m_{MSE}/(m_{MSE}+m_{DMS})$ from 0.37 to 1.0. As a crosslinking catalyst operating at room temperature [Bis(2-ethylhexanoate)tin] dissolved in 50 wt.-% dimethylpolysiloxan (SNB-1101, Gelest Inc. Morrisville, PA, USA) was added. The amount of catalyst was 1-2 wt.-% related to the tin metal. The liquid samples were poured in a polyethylen mold with a diameter of 10 mm and a hight of 8 mm for crosslinking. The as-processed samples were hold at 110 °C for 12 h and subsequently pyrolyzed in argon atmosphere at 1000 °C with a dwell time at maximum temperature of 2 h and a heating rate of 10 K/min, respectively. From the pyrolyzed samples the ceramic yield was calculated. The thermal transformation behavior was monitored by thermal analysis (TGA and DTA) with a simultaneously operating thermobalance STA 409A (Netzsch GmbH, Selb, Germany). About 50 mg of sample were placed in an alumina crucible and heated to 1000 °C in argon atmosphere with a heating rate of 10 K/min.

Viscosity measurements of the samples were carried out with a rotational rheometer (Haake VT 550, Thermo Electron GmbH, Karlsruhe, Germany) at 20 °C with shear rates of 10 and 100 s^{-1} at 20 °C. The viscosity adjustment was carried out with n-hexane, that was added to the MSE-100/DMS-S12 sample, which showed the highest ceramic yield after thermal conversion (sample with a MSE weight fraction of 0.7), in order to use the ink with an inkjet printer. The n-hexane volume fraction was varied from 0 to 0.26, related to the total volume fraction of the MSE-100/DMS-S12 sample. First printing experiments were carried out with a bubble jet printer of the type HP Deskjet 880C (Hewlett-Packard Company, Palo Alto, CA, USA). The color ink cartridge was opened by cutting the upper part with a band saw, removing the sponges from the three ink chambers for the cyan, magenta and yellow cartridge and cleaning the ink chambers with isopropanol by repeated flushing. A mixture of MSE-100/DMS-S12 with a $M_{MSE-100}=0.7$ was filled in one of the chambers and the catalyst which was delivered as a solution in polysiloxane was diluted in n-hexane and poured in another ink chamber. The composition for the first printing experimants was controled by a Cumputer aided design (CAD) software iGrafx DESIGNER Version 8.0.0512 (MICROGRAFX Inc., Richardson, Texas, USA). The pull-down menue for the **C**yan, **M**agenta und **Y**ellow color code for the subtractive color mixture allows the composition of each ink to be controlled from 0 to 100 by integer step. The chamber with the MSE-DMS ink was set to 100, and the chamber with the catalyst/n-hexane was set to 3-5. Printing was carried out first on paper (white inkjet paper, letter size, 80 g/m²), later on aluminum foil (0.01 mm in thickness; Reynolds Metals Company, Richmond, VA, USA), which was bonded to a sheet of paper.

RESULTS AND DISCUSSION

Crosslinking mechanism of the methoxymethyl(polysiloxane): Figure 1 shows the dynamic viscosity as a function of time of a MSE-100/DMS-S12 mixture (MSE-100 weight fraction=0.7) after catalyst addition. In this logarithmical scale, the viscosity increases slightly over a period of ~1700 seconds, and devolved into a step increase. When the temperature was increased prior to catalyst addition, the time of the linear viscosity increase could be reduced to about 200 seconds at 60 °C.

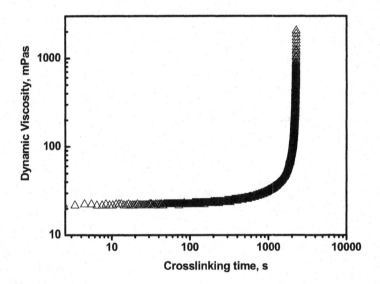

Figure 1.   Time-Viscosity dependence of an MSE-100/DMS-S12-mixture after catalyst adding (weight fraction $M_{MSE-100}$ = 0.7; Temperature=20 °C, Shear Rate=100s^{-1}).

Adjustment of the viscosity: The viscosity of the starting system with weight fraction of the siliconether $M_{MSE-100}$=0.7 was found to be 22.5 mPas at 20 °C, which is within the upper limit for inkjet printing. When fillers are introduced in the system, the viscosity is expected to increase. In order to keep the systems' dynamic viscosity below 30 mPas, which was shown to be the upper limit for inkjet printing in ref. [17], n-hexane was used for viscosity adjustment. The n-hexane shows no miscibility gap when mixed with the MSE-100/DMS-S12 system, has a viscosity of 0.31 mPas at room temperature and a boiling point of 69 °C, and thus allows rapid evaporation after printing. These physical properties make it a suitable modifier for the preceramic ink system. The resulting viscosity as a function of the n-hexane volume fraction is shown in Figure 2.

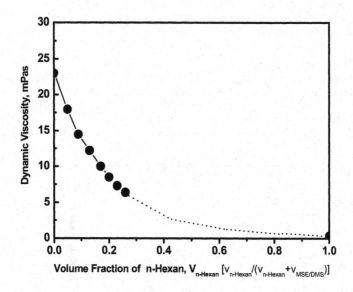

Figure 2.   Viscosity as a function of the n-hexane volume fraction at 20 °C.

An n-hexane volume fraction of only 0.05 decreased the viscosity to < 20 mPas, and the sample with a volume fraction of 0.26 showed a viscosity of 6.4 mPas.

Weight loss and ceramic yield: The TG curves of MSE-100/DMS-S12 mixtures with different MSE-100 weigt fractions are shown in Figure 3. The weight loss increased with increasing amount of MSE-100.

The weight loss after drying at 110 °C and the total weight loss after drying and pyrolysis at 1000 °C in argon atmosphere is shown in Figure 4. Even at a MSE-100 weight fraction of 0.37 a ceramic yield was detected. With an increasing weight fraction of MSE-100 the ceramic yield increased and showed a maximum at a weight fraction of 0.7 having a value of 54 wt. %. A further increase of weight fraction caused a decrease in the ceramic yield. From these findings it can be concluded, that fragments of the polysiloxane may influence the structure of the thermoset, and hence, increase the ceramic yield.

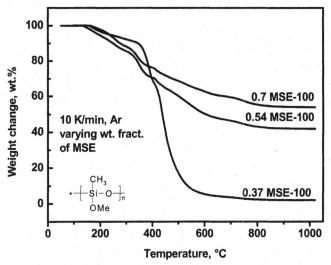

Figure 3. TG curves (top) and its first derivation (bottom) of MSE-100/DMS-12 mixtures with different weight fraction.

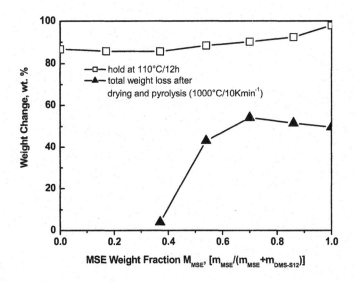

Figure 4. Weight change of the MSE-100/DMS-S12 samples with different MSE-100 weight fractions after drying at 100 °C and total weight change after drying and pyrolysis at 1000 °C in argon atmosphere.

CONCLUSIONS

A ceramic ink system for multichannel inkjet printing was developed based on a mixture of a liquid preceramic polymer and a latent water source. The water was generated *in situ* by the addition of a tin based catalyst. Both, the preceramic mixture and the catalyst can be added from different inkjet channels, and crosslink occurred at room temperature within minutes. The preceramic inks were investigated with respect to their ceramic yield after pyrolysis, and a ceramic residue of more than 50 wt. % was measured at optimized compositions. The preceramic ink can be diluted with n-hexane for viscosity adjustment which make them suitable for inkjet printing with particulate fillers.

Acknowledgement

The authors gratefully acknowledge the Deutsche Forschungsgemeinschaft DFG for financial support (SCHE628/3-1).

References

[1] H. Sirringhaus, T. Shimoda, "Inkjet printing of functional materials", *MRS Bulletin*, November 2003, 802-803.
[2] F.G. Zaugg, P. Wagner, "Drop-on-Demand Printing of Protein Biochip Arrays", *MRS Bulletin*, November 2003, 837-842.
[3] T. Shimoda, K. Morii, S. Seki, H. Kiguchi, "Inkjet Printing of Light-Emitting Polymer Displays", *MRS Bulletin*, November 2003, 821-827.
[4] S.E. Burns, P. Cain, J. Mills, J. Wang, H. Sirringhaus, "Inkjet Printing of Polymer Thin-Film Transistor Circuits", *MRS Bulletin*, November 2003, 829-834.
[5] C.X.F. Lam, X.M. Mo, S.H. Teoh, D.W. Hutmacher, "Scaffold development using 3D printing with a starch-based polymer", *Materials Science and Engineering* C **20** 49-56 (2000).
[6] H.M. Nur, J.H. Song, J.R.G. Evans, M.J. Edirisinghe, "Ink-jet printing of gold conductive tracks", *Journal of Materials Science: Materials in Electronics* **13** 213-219 (2002).
[7] H. Ago, J. Qi, K. Tsukagoshi, K. Murata, S. Ohshima, Y. Aoyagi, M. Yumura, "Catalytic growth of cabon nanotubes and their patterning based on ink-jet and lithographic techniques", *Journal of Electroanalytical Chemistry* **559** 25-30 (2003).
[8] J.R.G. Evans, M.J. Edirisinghe, P.V. Coveney, J. Eames, "Combinatorial Searches of Inorganic Materials using Ink-Jet Printer: Science, Philosophy and Technology", *Journal of the European Ceramic Society* **21** 2291-2299 (2001).
[9] X. Zhao, J.R.G. Evans, M.J. Edirisinghe, J.H. Song, "Ink-jet printing of ceramic pillar arrays", *Journal of Materials Science* **37** 1987-1992 (2002).
[10] M. Mott, J.R.G. Evans, "Solid Freeforming of Silicon Carbide by Inkjet Printing Using a Polymeric Precursor", *Journal of the American Ceramic Soc*iety **84** 307-313 (2001).
[11] A.R. Bhatti; M. Mott, J.R.G. Evans, M.J. Edirsinghe, "PZT pillars for 1-3 composites prepared by ink-jet printing", *Journal of Materials Science Letters* **20** 1245-1248 (2001).
[12] K.A.M. Seerden, N. Reis, J.R.G. Evans, P.S. Grant, J.W. Halloran, B. Derby, "Ink-jet printig of wax-based Alumina Suspensions", *Journal of the American Ceramic Soc*iety **84** 2514-2520 (2001).
[13] M. Mott, J.R.G. Evans, "Solid Freeform of Silicon Carbide by Inkjet Printing Using a Polymeric Precursor", *Journal of the American Ceramic Soc*iety **84** 307-313 (2001).

[14] D. Seyferth, N. Bryson, D.P. Workman, C.A. Sobon, Preceramic Polymers as 'Reagents' in the Preparation of Ceramics, *Journal of the American Ceramic Society* **74** 2687-2689 (1991).

[15] P. Greil, "Active Filler Controlled Pyrolysis of Preceramic Polymers (AFCOP)", *Journal of the American Ceramic Society* **78** 835-848 (1995).

[16] M. Stackpoole, R.K. Bordia, "Reactive Processing and Mechanical Properties of Si_3N_4 Matrix Composites", *Ceramic Transactions* **108** 111-119 (2000).

[17] D.H. Lee, B. Derby, Preparation of PZT suspensions for direct ink jet printing *Journal of the European Ceramic Society* **24** 1069-1072 (2004).

Plasma Synthesis

EFFECT OF PARTICLE SIZE ON THE CHARACTERISTICS OF MULLITE-ZrO$_2$ CERAMICS PRODUCED BY SPARK PLASMA REACTION SINTERING

Enrique Rocha-Rangel
Dept. of Materials, Universidad Autónoma Metropolitana, México, D.F. 02200

Sebastián Díaz de la Torre
CIMAV Chihuahua, 31109 México

Heberto Balmori-Ramírez
Dept. of Metallurgical and Materials Eng., National Polytechnic Institute,
A.P. 118-389, México, D.F., 07300

ABSTRACT

Zirconia-mullite composites were produced by spark plasma reaction sintering (SPRS) with a powder mixture of ZrSiO$_4$ + Al$_2$O$_3$ + Al. The powder was milled in an attritor for 6 and 12 h in order to prepare two different particle sizes to evaluate the influence of the particle size on the formation of mullite and densification. The attrition-milled powders were calcined at 1100°C before the SPRS. They were consolidated at different temperatures up to 1580°C with a pressure of 40 MPa. The progress of the mullitization reaction and the densification was faster for the composites fabricated with the finer particles. The microstructure of the composites fabricated with the finer particles achieved almost full density and were completely mullitized. Their microstructure consisted of a homogeneous dispersion of tetragonal and monoclinic ZrO$_2$ particles in a mullite matrix. They achieved a hardness of 1533 MPa and a high Young's modulus. The flexural strength and toughness were 797 MPa and 5.93 MPa·m$^{1/2}$, respectively.

INTRODUCTION

Mullite has favorable physical and chemical properties such as excellent resistance to thermal shock and chemical stability. However, its applications are somewhat limited because of poor toughness and strength [1]. The mechanical properties of mullite are improved by incorporating ZrO$_2$ particles dispersed in the matrix [2-4]. Mullite-zirconia composites can be fabricated by a reaction-sintering process with zircon (ZrSiO$_4$) and alumina (Al$_2$O$_3$) according to the following reaction [5-8]:

$$3Al_2O_3 + 2ZrSiO_4 \rightarrow 3Al_2O_3 \cdot 2SiO_2 + 2ZrO_2 \qquad (1)$$

Spark Plasma Sintering (SPS) has been used to consolidate ceramics and composites in short sintering times and low temperatures [8, 9]. It has also been used to consolidate mullite/ZrO_2 ceramics [10]. The advantage of using SPS lies in the fact that it can provide rapid heating of the whole sample, allowing low sintering temperatures and short sintering times. Homogeneous sintered bodies can be obtained [5, 6]. Employing SPS process, one can pass quickly through the surface diffusion controlled regime where to the boundary or lattice-diffusion-controlled regimes where densification mechanisms operate.

In this paper, the phase evolution, microstructure development and mechanical properties of mullite/ZrO_2 ceramics, starting from mixtures of Al_2O_3 and $ZrSiO_4$ powders milled in an attritor for different times (6 and 12 h), were studied. The effect of the initial particle size of the powders on the characteristics of the SPS sintered bodies will be discussed.

EXPERIMENTAL PROCEDURE

Batches (100 g) of zircon (~1µm, Kreutz, Germany), aluminum (~5µm, Analytical de Mexico) and α-alumina (TM-10, Taimei, Japan), mixed in a ratio of 64/18.5/17.5 wt %, were prepared. The powders were attrition milled in isopropyl alcohol for 6 and 12 h at 400 rpm in a Union Process mill device with 3 kg of YTZ-Zirconia balls of 3 mm diameter. The attrition milled powders were heated up to 1100°C at 1°C/min in air to ensure the oxidation of the aluminum metal. The sum of the α-Al_2O_3 and the alumina formed in-situ after the oxidation of the aluminum corresponds to the stoichiometric amount of alumina required by reaction (1). The calcined powder was milled in a planetary mill for 30 min to break the agglomerates that were formed during the calcination. The specific surface area (BET) of the powders was evaluated with an automatic volumetric gas sorption analyzer (AUTOSORB-1, Quanta Chrome, USA). The composites prepared with powders milled for 6 and 12 h are going to be identified as R6 and R12 respectively.

Small batches (3 g) of the attrition milled, calcined and planetary ground powders R6 and R12 were packed in the cylindrical graphite die (20 mm inner diameter x 40 mm long) of the SPS machine (Dr. Sinter-SPS 1020, Sumitomo Coal Mining Co., Japan). Sintering temperatures were changed between 1400°C and 1580°C for the R6 powders, and between 1350°C and 1640°C for the R12 powders. Major details of the SPS process conditions have been presented elsewhere [12].

The density and porosity of the consolidated pellets were evaluated by the Archimedes principle. The phases and microstructure of the sintered samples was analyzed by X-ray diffraction (XRD, Siemens D-5000, Germany) and scanning electron microscopy (SEM, Jeol-6300, Japan) respectively, with samples polished with standard ceramographic techniques. The Young's modulus (E) was measured with an impulse-resonance method (Grind-O-Sonic). The fracture toughness (K_{IC}) was measured by indentation applying a load of 50 kg [13].

RESULTS AND DISCUSSION

Specific surface area

The specific surface area of the mixture of zircon, alumina and aluminum powders before milling was of 2.96 m^2g^{-1}. The specific surface areas after attrition millng and after the calcination at 1100°C and grinding in the planetary mill are shown in Table 1. Aluminum oxidation causes a weight gain of 8.39% and 6.70% in the samples R6 and R12, respectively. The higher specific surface area of the powder R12 indicates that it is finer and more reactive than the powder R6.

Table 1. Specific surface area of the mixture of zircon, alumina and aluminum powders.

Milling time	After attrition milling	After calcinations at 1100°C and grinding in the planetary mill
6 h	16.46 (m^2/g)	36.38 (m^2/g)
12 h	21.57 (m^2/g)	52.57 (m^2/g)

X-ray Diffraction

Figure 1 shows the XRD patterns of the R12 composites at different stages of processing. The XRD pattern of the as-received powder has the peaks of $ZrSiO_4$ (Z), Al (ϕ) and α-Al_2O_3 (α). The powder calcined at 1100°C shows peaks of $ZrSiO_4$, α-Al_2O_3 and two diffuse peaks at 2θ = 37.5° and 45° that correspond to γ-Al_2O_3 (γ). The absence of the peaks of Al in this pattern indicates that this metal, or most of it, has been oxidized during the calcination at 1100°C. The XRD pattern of the composites fired at 1350° has peaks of $ZrSiO_4$, α-Al_2O_3 and small peaks of tetragonal ZrO_2. It means that the γ-Al_2O_3 transformed to α-Al_2O_3. The ZrO_2 comes from the decomposition of zircon. At 1420°C, the XRD intensities of the peaks of zircon and α-Al_2O_3 have decreased with respect to the intensity of the same peaks which are observed in the XRD pattern of the composite fired at 1420°C. The peaks of these

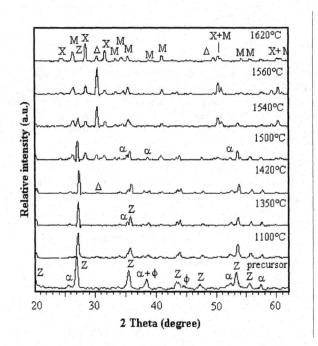

Figure 1. X-Ray diffraction patterns of milled powder for 12 h and of samples sintered by SPRS at different temperatures. Z:ZrSiO$_4$, M:Mullite, ϕ:Al, α:α-Al$_2$O$_3$, X:m-ZrO$_2$, Δ:t-ZrO$_2$.

two phases disappear at 1560°C and 1500°C respectively. Traces corresponding to mullite and t-ZrO$_2$ are observed in the pattern at 1420°C. The intensity of the peaks of these phases increases as the firing temperature increases. In the composites fired at 1500°C, ZrO$_2$ is present in both its monoclinic and tetragonal phases. The formation of mullite and the decomposition of ZrSiO$_4$ are completed at 1560°C.

The progress of reaction (1) for the composites R6 was also followed with XRD. A comparison of the formation of mullite for the two composites R6 and R12 was made by calculating the relative intensities of the main peaks of mullite and zircon, using the equation (2):

$$Mullitization = \frac{I_{210}^{M}}{I_{210}^{M} + I_{200}^{Z}} \qquad (2)$$

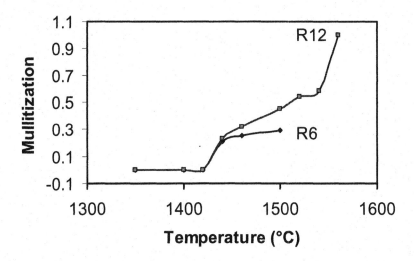

Figure 2. Influence of the SPRS temperature on the formation of mullite in the two composites.

Where I^M_{210} is the integrated intensity of the (210) peak of mullite, and I^Z_{200} is the (200) peak of zircon. The formation of mullite at different temperatures, as determined with equation (2), is shown in Figure (2). The first traces of the decomposition of zircon and of the formation of mullite were observed at 1420°C for both powders, R6 and R12. The progress of the reaction (1) is similar until a temperature of 1440°C is reached. From thereafter, the formation of mullite is slower in the composites R6 than in the composites R12.

Densification

The relative densities change of the two composites R6 and R12, as a function of the SPRS temperature is presented in Figure 3. This figure shows that the composite R12 which has a smaller particle size, densifies faster than the powder R6. The relative densities of the composites R6 and R12 fired at 1400°C in the SPS machine were 92 % and 95.5 %, respectively. The density of the two composites increased continuously as the firing temperature increased, but the composites R12 were always denser than the composites R6. At a sintering temperature of 1500°C, the composites R6 had a relative density of 95 %, while the composites R12 had a relative density of 99.1 %. Measurements of the amount of open porosity of the composites fired at 1500°C showed that the composites R6 had 0.7 % open porosity and that the composite R12 had only 0.2 %. The composites R6 were not fired at higher temperatures. The composites R12 achieved almost full density at a firing temperature of 1560°C.

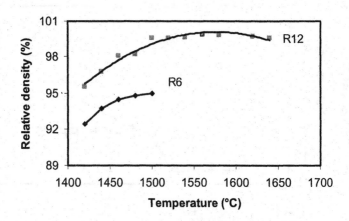

Figure 3. Effect of sintered temperature on relative density
of samples R6 and R12 sintered by SPRS.

Microstructure

The microstructure of the composite R6 fired at different temperatures is shown in Figure 4. The sample R6 fired at 1410°C contains three constituents (Figure 4a). They were identified with an EDX analysis in the SEM. The light gray phase was identified as zircon. The white bright particles have been identified as zirconia. The matrix contains a mixture of both alumina and mullite. The size of the zircon particles is greater than 1 µm. The sample fired at 1500°C also has three constituents, which have been identified as zircon (light gray particles), zirconia (bright particles) and mullite (darker matrix). This composite contained defects like those shown in Figure 4c. The defects more commonly found were big particles of zircon, residual porosity and crack-like defects that were probably originated from agglomerates.

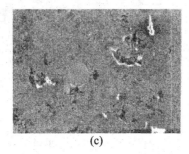

$\overline{1\ \mu m}$

(a)   (b)

$\overline{10\ \mu m}$

(c)

Figure 4. Microstructures of sample R6 sintered at different temperatures
a) 1410°C, b) 1500°C and c) Defects in samples R6.

    The microstructure of the composites R12 fired at 1420°C and 1500°C are shown in Figure 5a and b, respectively. These microstructures have the same microconstituents that were identified in the composites R6, but their size is finer. The composite fired at 1560°C contained only a mixture of zirconia and mullite (Figure 5c).

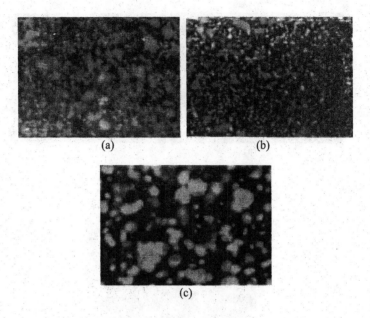

1 μm

(a)   (b)

(c)

Figure 5. Microstructures of sample R12 sintered at different temperatures
b) 1420°C, b) 1500°C and c) 1560°C.

Mechanical properties

Table II shows the results of the mechanical properties measured in the samples R6 and R12 as a function of the sintering temperature. In general, both samples have good mechanical properties. The fracture toughness of the samples increases with the temperature. The fracture toughness reached a maximum value of 5.97 MPa·m$^{1/2}$ in the sample R12 sintered at 1560°C. This is a higher fracture toughness than for mullite-zirconia composites processed by conventional methods, which typically have 2-4 MPa·m$^{1/2}$ [2, 4].

The improvement in the fracture toughness is attributed to the high density and homogeneous microstructure of the composities processed by the SPRS technique. Because the t-ZrO_2 content in these composites is high, the toughening mechanism that operated in them is the zirconia transformation mechanism.

The fracture strength of the composites was measured by three point bending. Only a limited number of samples (3) were tested at some temperatures. Therefore, these results can only be taken as a guidance of the properties that can be attained by SPRS. In general, the composites R12 had higher strengths than the composites R12. Of particular relevance is the value of 797 MPa that was obtained for the composite R12 fired at 1580°C.

Table I. Mechanical properties of the mullite-ZrO_2 composites produced by SPRS.

Sintering temp. (°C)	Sample R6				Sample R12			
	HV (GPa)	E (GPa)	K_{IC} (MPam$^{1/2}$)	σ (MPa)	HV (GPa)	E (GPa)	K_{IC} (MPam$^{1/2}$)	σ (MPa)
1420	1790	179	2.87	---	1865	226	3.15	74.5
1460	1396	200	3.10	---	1530	193	2.93	707
1480	1394	192	2.41	364	1506	183	3.0	323
1500	1436	234	3.42	410	1515	207	3.20	390
1580	---	---	---	---	1353	187	5.93	797

HV: hardness, E:Young's modulus, K_{IC}: Fracture toughness, σ: Strength

CONCLUSIONS

Mullite-zirconia composites can be fabricated by spark plasma reaction sintering (SPRS) with a mixture of alumina and zircon. Finer particles were produced by attrition milling for longer times. The analysis of the effect of the starting size particle on the progress of the mullitization reaction, densification, microstructure and mechanical properties of mullite-ZrO_2 composites shows that finer particles lead to faster reaction, higher densification and better mechanical properties.

REFERENCES

1. J. S. Moya and M. I. Osendi, "Microstructure and Mechanical Properties of Mullite/ZrO_2 Composites", *J. Mat. Sci.*, 19 (1984)2909-2914.
2. N. Claussen and J. Jahn, "Mechanical Properties of Sintered In Situ Reacted Mullite Zirconia Composites", *J. Am. Ceram. Soc.*, 63(1980) 228-229.
3. K. Rundgren, P. Elfing, H. Tabata, S. Kansaki and R. Pompe, "Microstructure and Mechanical Properties of Mullite-Zirconia Composites Made from Inorganic Sols and Salts", pp. 553-566 in *Ceram. Trans. Vol. 6: Mullite and Mullite Matrix Composites*, ed. by S. Somiya, R. F. Davis and J. A. Pask, American Ceramic Society, Westerville, OH, 1990.
4. T. Koyama, S. Hayashi, A. Yasumori and K. Okada, "Contribution of Microstructure to The Toughness of Mullite/Zirconia Composites", pp. 695-700 in *Ceramic Transactions Vol. 51*, ed. (1995).

5. H. Schneider, K. Okada and J. A. Pask, *Mullite and Mullite Ceramics*, John Wiley & Sons, Inc., New York, 1994.

6. T. Koyama, S. Hayashi, A. Yasumori and K. Okada, "Preparation and Characterization of Mullite-Zirconia Composites from Various Starting Materials", *J. Eur. Ceram. Soc.*, 14, (1994)295-302.]

7. K. A. Khor and Y. Li, "Effects of Mechanical Alloying on the Reaction Sintering of $ZrSiO_4$ and Al_2O_3", *Mater. Sci. Eng.* A256 (1998)271–279.

8. J. R. Groza, M. Garcia and J. A. Schneider, "Surface Effects in Field-Assisted Sintering", *J. Mater. Res.*, 16[1]286-292(2001).

9. S. D. de la Torre, H. Miyamoto, K. Miyamoto, J. Hong, L. Gao, L. Tinoco-D., E. Rocha-Rangel and H. Balmori-Ramirez, "Spark Plasma Sintering of Nano-Composite Ceramics", pp. 892-897 in *Proc. 6th International Symposium on Ceramic Materials and Components for Engineering*, Arita, Japan, 1997.

10. K. A. Khor, L. G. Yu, Y. Li, Z. L. Dong, Z. A. Munir, "Spark Plasma Reaction Sintering of ZrO_2-Mullite Composites From Plasma Spheroidized Zircon/Alumina Powders", *Mater. Sci. Eng.* A339(2003)286-296.

11. J. A. Schneider, R. S. Mishra and A. K. Mukherjee, "Plasma Activated Sintering of Ceramic Materials", pp. 143-151 in *Ceram. Trans. vol. 79: Advanced Synthesis and Processing of Composites and Advanced Ceramics II*, ed. by K. V. Logan, Z. A. Munir and R. M. Spriggs, American Ceramic Society, Westerville OH, 1996.

12. E. Rocha-Rangel, H. Balmori-Ramírez and S.D. De la Torre, "Reaction-Bonded Mullite-ZrO_2 Composites with Oxide Additives", pp 231-236 in Ceramic Transactions vol. 96, Advances in Ceramic Matrix Composites IV, ed. by J.P. Singh and N.P. Bansal The American Ceramic Society, Ohio, USA 1999.

13. K. Niihara, R. Morena and D. P. H. Hasselman, "Evaluation of K_{IC} of Brittle Solids by the Indentation Method with Low Crack-To-Indent Ratios", *J. Mater. Sci. Lett.*, 1(1982)13-16.

Composites

SYNTHESIS AND CHARACTERIZATION OF Al_2O_3-SiC COMPOSITE POWDERS FROM CARBON COATED PRECURSORS

Soydan Ozcan and Rasit Koc
Department of Mechanical Engineering and Energy Processes
Southern Illinois University
Carbondale, Illinois 62901-6603

ABSTRACT

The synthesis of submicron Al_2O_3-SiC composite powders from carbon coated SiO_2/Al mixture was investigated. Two different types of precursors which are carbon coated SiO_2/Al mixture and standard mixture SiO_2, Al and carbon black were prepared to compare formation of the composite powders. The effect of temperature (1200-1500°C) on the formation of the powders was studied in flowing inert gas. Al_2O_3-SiC composites were characterized using X-ray diffraction (XRD), Brunauer-Emmett-Teller (BET) surface area analyzer and transmission electron microscopy (TEM). The coating of SiO_2/Al mixture significantly changed the mechanism of the reaction and produced high quality nano-size Al_2O_3-SiC powders. XRD and BET results showed that precursors synthesized at 1300°C for 2 hours had only Al_2O_3 and SiC phases and had high surface area about $36m^2/g$, respectively. TEM results showed that the produced Al_2O_3-SiC powders had fine particles, narrow particle distribution, and freely agglomerated.

*Sponsored by the U.S. Department of Energy, Office of Industrial Technologies, and Industrial Materials for the Future.

INTRODUCTION

Advanced ceramics and composites play an important role in the industries of the future. The development and use of advanced ceramics are revolutionizing the field of material science and technology. Dramatic changes are taking place in the way materials are used and fabricated. The results of this foment are the improved performance of existing products and the birth of new technologies [1]. Among of those advanced ceramics, Al_2O_3-SiC composites have been used as cutting tools and forming dies because of their high wear resistance, strength and fracture toughness [2-4]. The incorporation of SiC into Al_2O_3 matrix results into significant increase in fracture toughness, the level of toughening being strongly dependent on morphology of SiC [5,4]. The adding of SiC to Al_2O_3 can limit the matrix grain growth during sintering and increase the mechanical properties of resultant composites. Recently, Niihara et al [6] and Young-Keun et al [7] reported that the incorporation of nanosize dispersoids into a ceramic matrix can lead significant improvements in mechanical properties, since then there has been a growing interest in ceramic-matrix nano-composites. For the improved mechanical properties, the Al_2O_3 powders with a homogenous distribution of SiC particles, fine particle size, a narrow particle size distribution, and a loose agglomeration are required.

It has been shown that agglomeration in powder due to the small particle size is very serious, resulting in a heterogeneous size distribution of particles in the matrix when Al_2O_3 and SiC is conventionally mixed. It is therefore necessary to find a new method to synthesize the Al_2O_3-SiC powders with homogenous size distribution.

The carbon coating method [8-10] developed under Dr. Koc's currently funded DOE-OIT Advanced Industrial Materials-Industrial Materials of the Future Program has resulted into the development of two US patents (US Patent No: 5,417,952 [11] deals with the production of nanosize β-SiC powders and this process received R&D 100 in 1995; US Patent No: 5,342,494 [12], deals with the synthesis of nano size TiC powders).

In this paper, the combination of the carbon coating method and SHS reaction technique is investigated to produce Al$_2$O$_3$-SiC nano composites. The emphasis of the investigation is to quantify the effects of reaction conditions on the composition, size shape and crystalinity.

EXPERIMENTAL PROCEDURE

For this study, two different types of precursors which are carbon coated SiO$_2$/Al and standard mixture SiO$_2$, Al and carbon black were prepared. SiO$_2$ (P-25, Degussa Corp., Ridgefield Park, NJ), Al (41000, Alfa Aesar, Ward Hill, MA) and carbon black (Monarch 880, Cobat, Waltham, MA) were used as a starting materials.

Preparation of carbon coated SiO$_2$/Al mixtures (fully coated):

Carbon coated SiO$_2$/Al mixtures produced using the method that was first described by Koc et al. [11]. Desired weight percent ratios were thoroughly mixed in ethyl alcohol using an attritor (Model HD01-A100-9, Union Process, Akron, OH). After drying with a vacuum oven (Model 1400E, VWR Scientific, Philadelphia, PA), the mixtures were milled by a Spex, Mixer/Mill (Model 8000, Spex, Metuchen, NJ). Then, a rotating coating apparatus, consisting of 10 cm ID x 35 cm long stainless steel vessel, was used for preparing low density, pyrolitic-carbon coated mixture particles utilizing propylene gas (C$_3$H$_6$) as the coating gas. About 200g of SiO$_2$/Al mixture was placed in the vessel and the vessel was evacuated to a moderate vacuum level using a rotary vacuum pump. Then, the vessel was filled with propylene gas until the pressure reached 3.0 atm. The vessel was then heated to 550°C, while it was rotating. The coating step was continued until the 11 wt% carbon was deposited.

Preparation of SiO$_2$, Al and carbon black mixtures (Mixed):

Desired weight percent ratios of SiO$_2$, Al and carbon black were first mixed in ethyl alcohol by an attritor mill for 4 hours. The ratio of the mixture and ethyl alcohol was 1:0.5 by volume. The mixed powders were then dried in a vacuum oven.

Synthesis of Al$_2$O$_3$-SiC from carbon coated SiO$_2$/Al mixtures and SiO$_2$, Al and carbon black mixtures:

Ten grams of each precursors were placed in graphite crucibles and synthesized at temperatures of 1200, 1300, 1400 and 1500°C for 2 hours in a box furnace to investigate reaction mechanism and determine an optimal reaction condition. Heating and cooling rates of 4°C/min were used under the flowing argon gas condition.

The produced powders were characterized using X-Ray diffraction (XRD) (Rigaku, Tokyo, Japan) with Cu$_{K\alpha}$ radiation, a BET surface area analyzer (Micromeritics, Gemini 2360, Norcros, GA), and transmission electron microscopy (TEM) (Hitachi, Model FA 7100, Tokyo, Japan).

RESULTS AND DISCUSSION

Figure 1 shows the XRD patterns obtained at temperatures in the range 1200-1500°C from the carbon coated precursors. At 1400°C, formation of Al_2O_3-SiC composite powders was completed. BET measurements showed that powders had a surface area of 26 m^2/g (See Table 1). High temperature DSC studies will be conducted to provide information on the formation mechanism for Al_2O_3-SiC from carbon coated SiO_2/Al mixtures.

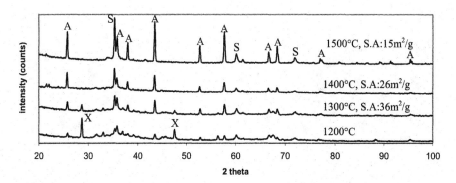

Fig. 1 XRD patterns of coated precursors synthesized at various temperatures. A=Al_2O_3, S=SiC, X: crystalline silica. (S.A: Surface Area)

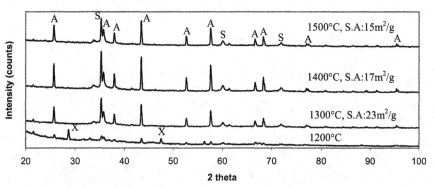

Fig. 2 XRD patterns of mixed precursors synthesized at various temperatures. A=Al_2O_3, S=SiC, X: crystalline silica. (S.A: Surface area)

Figure 2 shows the XRD patterns obtained at temperatures in the range 1200-1500°C from conventionally mixed powders. Similar to coated precursors, at 1400°C, Al_2O_3-SiC formation was completed. However, the BET surface areas of resulting powders were less that of the powders produced from carbon coated precursors (See Table 1).

Table I. BET Surface area results of products synthesized at different temperature

| Synthesis Temperature (°C) | Surface Area (m²/g) | |
	Coated Precursors	Mixed Precursors
1300	36	23
1400	26	17
1500	15	15

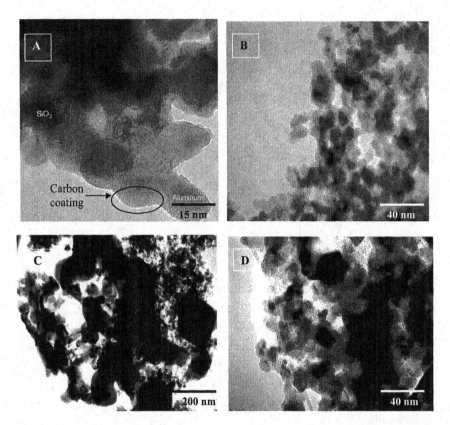

Fig. 3 TEM micrographs of synthesized products at 1400 °C for 2 hours in flowing argon a,b) coated c,d) mixed precursors.

Figure 3a and b shows the carbon coated precursor and powder morphology of resulting from the precursors, respectively. These TEM micrographs indicate that the thickness of carbon coating on the starting powders was in nanosize and the particles of final product was not agglomerated. Figure 3c and 3d is the TEM micrographs of conventionally mixed precursor and the final product from the mixed powders. Figure 3d shows the final powder morphology where the particles appear to be agglomerated. These results are from our early investigations. We attempt to provide information on the advantages of the carbon coated precursors. The kinetics studies combined with XRD and TEM will be reported in upcoming publications.

CONCLUSION

The production of Al_2O_3-SiC was investigated using two different precursors. The carbon coating precursor showed that the products had only two phases of Al_2O_3 and SiC with very high surface area. The conventionally mixed precursors showed the synthesized powder under similar conditions had very low surface area with hard agglomerates. These initial results indicate that the carbon coated SiO_2/Al mixture leads to an intimate mixing of the reactant and can be very effective way for controlling reaction path and rate in SHS reactions.

REFERENCES

[1]A. Krell and P. Blank, TiC-strengthened Al_2O_3 by Powder Tailoring and Doping Procedures *Materials Science and Engineering A*, **161** 295-301 (1993).

[2]J. Zhao, L. Stearns, M. Harmer, H.M. Chan, G.A. Miller, and R.E. Cook, "Mechanical Behavior of Alumina-Silicon Carbide Nanocomposites", *Journal of American Ceramic Society*, **76** [2] 503-10 (1993).

[3]E. Liden, E. Carlstrom, L. Eklund, B. Nyberg, and R. Carlson, "Homogenous Distribution of Sintering Additives in Liquid-Phase Sintered Silicon Carbide", *Journal of American Ceramic Society*, **78** [7] 1761-68 (1995).

[4]M. Sternitzke, B. Derby, and R.J. Brook, "Alumina/Silicon Carbide Nanocomposites by Hybrid Polymer/Powder Processing: Microstructures and Mechanical Properties", *Journal of American Ceramic Society*, 81 [1] 41-48 (1998).

[5]Q. Yang and T. Troczynski, "Alumina Sol-Assisted Sintering of SiC-Al_2O_3 Composites", *Journal of American Ceramic Society,* **83** [4] 958-60 (2000).

[6]K. Niihara, "New Design Concept of Structural Ceramics-Ceramics Nanocomposites", *Journal of Ceramic Society Japan.*, **99** [10] 972-82 (1991).

[7]Y. Jeong, A. Nakahira, K. Niihara, "Effects of Additives on Microstructure and Properties of Alumina-Silicon Carbide Nanocomposites", *Journal of American Ceramic Society,* **82** [12] 3609-12 (1999).

[8]G. A. Swift and R. Koc, "Tungsten Powder from Carbon Coated WO_3 Precursors", *Journal of Material Science*, **36** [4], 803-6 (2001).

[9]R. Koc, G. Glatzmaier and J. Sibold, "β-SiC Production By Reacting Silica Gel With Hydrocarbon Gas", *Journal of Material Science*, **36** [4], 995-99 (2001).

[10]R. Koc, C. Meng and G.A. Swift, "Sintering Properties of Submicron TiC Powders From Carbon Coated Titania Precursor", *Journal of Material Science*, **35** [12], 3131-41 (2000).

[11]G. Glatzmaier and R. Koc, "Method for Silicon Carbide Production by Reacting Silica with Hydrocarbon Gas", US Patent No: 5,324,494 (1994).

[12]R. Koc and G. Glatzmaier, "Method for Synthesizing TiC, TiN and Ti(C,N) Powders", US Patent No: 5,417,952 (1995).

CELLULAR β-SIALON/SIC COMPOSITE CERAMICS FROM CARDBOARD

C.R. Rambo and H. Sieber

University of Erlangen-Nuremberg, Department of Materials Science, Glass and Ceramics, Martensstr. 5, D-91058, Erlangen, Germany.

ABSTRACT

Lightweight cellular β-SiAlON / SiC ceramics were produced via infiltration (dip coating at ambient temperatures) of metal powder / preceramic polymer containing slurries into preformed corrugated cardboard templates. After infiltration, the templates were pyrolysed in Ar atmosphere at 800°C and 1200°C to decompose the cellulose into carbon and to promote the infiltration and reaction with the molten Si/Al-slurry into SiC. Subsequently the porous templates were nitrided at 1200°C - 1530°C resulting in the formation of a $Si_{6-z}Al_zO_zN_{8-z}$/ SiC composite with z varying from 0.6 to 1.0.

INTRODUCTION

SiAlON-based ceramics and ceramic composites are potential candidates for high temperature engineering applications [1,2]. The mechanical properties of SiAlONs are similar to Si_3N_4 but with a superior oxidation and creep resistance. A low density and coefficient of thermal expansion, even lower than for other nitride and oxide ceramics [3], are highly attractive for packaging materials in microelectronics or components for aeronautics. Table 1 summarizes some characteristic properties of SiAlONs compared to Al_2O_3 and Si_3N_4.

Tab. 1: *Physical properties of Al_2O_3, Si_3N_4 (Reaction bonded silicon nitride-RSBN) and SiAlON.*

		Al_2O_3	Si_3N_4	SiAlON
Density	(g/cm^3)	3.96	3.2	3.08-3.18
Elastic Modulus	(GPa)	386	250-320	300
Coefficient of Thermal Expansion	(x10^{-8}/°C)	7.4	3.0	1.9-3.3

Two types of SiAlON are of interest for engineering applications,: β-SiAlON (β′) and α-SiAlON (α′). The term SiAlON denotes solid solutions of Si_3N_4 in the β-type modification, β- $Si_{6-z}Al_zO_zN_{8-z}$, (z = 0-4.2) or α-$M_xSi_{12-(m+n)}Al_{(m+n)}O_nN_{16-n}$ where x is the amount of stabilizing cation M (M=Li^+, Mg^{2+}, Ca^{2+}, Y^{3+} and lanthanides with z ≥ 60). x may vary from 0.3-0.5 with n = 0.75-1.26 and m = 1-1.5, respectively or mixtures thereof [4]. Since the last two decades a variety of synthesis routes and raw materials were reported for the preparation of SiAlON ceramics [5-8]. Beyond the studies of pure β′ phase, SiAlON-SiC composite ceramics were investigated mainly focusing the improvement of the mechanical properties of the composites, specially at high temperatures [9-11].

The processing of light-weight, porous ceramics and ceramic composites from natural cellulose fiber materials has attained increasing attention in the recent decade. Low cost raw materials as well as availability of well established paper processing offer an economic way to manufacture lightweight ceramic composites [12,13]. Previous work was focused on the conversion of cardboard structures into SiC-based ceramics and SiC / Si-Al-O-C ceramic composites [14]. Dip coating with Al/Si-containing slurries followed by reaction of the Al-Si above the melting temperature with the carbonized perform resulted in SiC-Al$_2$O$_3$-Mullite ceramics. In contrast to sintering of a porous powder preform, the cardboard preforms showed only minor little shrinkage, which facilitates near-net shape manufacturing of large light-weight devices such as panels or corrugated structures [15]. The aim of the present work was the manufacturing of light-weight SiAlON / SiC ceramic composites from preformed cellulose fiber cardboard structures. A Si/Al/poly-methylsiloxane slurry was selected as a source for infiltration and reaction with the carbonized perform which finally resulted in β-Sialon containing composites.

EXPERIMENTAL

Commercially available corrugated cardboard made out of cellulose fibers was used as template. A corrugated cardboard (corrugation area weight of 125 g/m^2 and top layer area weight of 165 g/m^2) with a cell diameter between 2 and 6 mm and a cell wall thickness of 100 μm (*190TL, Wellpappenwerk, Bruchsal / Germany*) was pre-shaped in cellular cylinders of about 2 cm of height and 3.5 cm in diameter, dried at 70°C and infiltrated by dip coating with a metal powder-polymer slurry. The slurry was prepared with isopropyl alcohol containing 40 vol.% Si-powder, 40 vol.% Al-powder and 20 vol.% of a polysilsesquioxane polymer powder $(CH_3SiO_{1.5})_n$, with n = 300-400 (NH21 - poly(methylsiloxane), *Hüls Silikone GmbH, Nünchritz / Germany*). The mean particle size of the metallic powders was below 10 μm for Si and 50 μm for Al. The Al/Si ratio corresponds to a weight fraction of 53.8 wt% Al and 46.2 wt% Si, which exhibits a melting temperature of 1012°C.

After the dip coating, the specimens were pyrolysed in inert atmosphere at 800°C and 1200°C for 1 h to decompose the cellulose based fibers of the cardboard into carbon and the preceramic polymer into an amorphous Si-O-C residue [16]. At temperatures above 1012°C the Si/Al-powder slurry melts and infiltrates the pyrolysed carbon fiber template structure. The pyrolysis was carried out under flowing Ar, in order to prevent the formation of nitride phases that could avoid the Al/Si melt-infiltration into the carbon at temperatures below the Al/Si melting temperature. After pyrolysis the samples were submitted to nitridation. The nitridation was performed in a tubular furnace at temperatures between 1200-1530°C in flowing N$_2$-atmosphere for 3 h. Fig. 1 shows the schematic processing route for manufacturing of cellular β-SiAlON/SiC ceramics from corrugated cardboard.

The phase evaluation during the nitridation process was analyzed by X-ray diffractometry (XRD) (*D 500, Siemens, Karlsruhe / Germany*). The z values of the obtained β-SiAlON phase were calculated using the relationships between z and the unit-cell dimensions a$_o$ and c$_o$ of the β-Si$_3$N$_4$ (hexagonal, space group P6$_3$) according to *K.H. Jack* [4]. The microstructures of the cellular ceramic composites were characterized by scanning electron microscopy (SEM) (*Phillips XL 30, Nederland*). The skeleton density was measured by He-pycnometry (*Accu Pyk 1330, Micromeritics, Düsseldorf / Germany*).

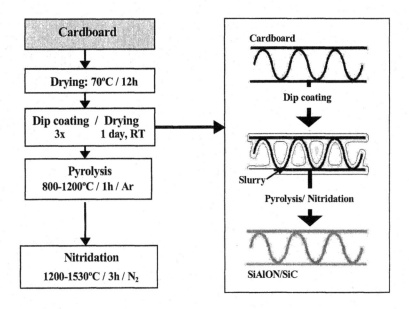

Figure 1: *Schematic diagram of the processing route of cellular β-SiAlON/SiC ceramics from corrugated cardboard.*

RESULTS AND DISCUSSION

The linear shrinkage of the dip-coated cardboard specimen after pyrolysis and ceramic phase reaction was about 3% in the radial and axial directions. It is mainly related to the large weight loss of the cellulose fibres (between 60-80 wt.%) during pyrolysis [14]. After final nitridation no further shrinkage could be detected.

Fig. 2 shows the XRD-spectra of the specimens after pyrolysis at 800°C and 1200°C and after final nitridation at temperatures between 1200-1530°C. After pyrolysis at 800°C only elemental Al and Si were detected. Pyrolysis at 1200°C in Ar yielded the formation of SiC and Al_2O_3. The oxygen necessary for the oxidation of the Al was provided by the preceramic polymer, which decomposed into an amorphous Si-O-C phase after pyrolysis at 800°C and into β-SiC, SiO_2 and carbon above 1300°C [16]. During the infiltration process, Al partially reduced the SiO_2 leading to free Si and Al_2O_3. A temperature of 1200°C was determined to be the minimum temperature at which the Al/Si-melt sufficiently can wet and infiltrate into the porous carbon template.

When nitridation was carried out, AlN was formed as the first nitride phase. At 1430°C a substantial amount of β-Si_3N_4 was formed. β-$Si_{6-z}Al_zO_zN_{8-z}$ formation was found in literature to require temperatures exceeding 1250°C with the maximum fraction being formed at temperatures between 1450°C and 1550°C, depending on the starting fraction of AlN, Al_2O_3 and Si_3N_4 [17].

After nitridation at 1430°C for 3 h, the formation of a small amount of the β-SiAlON-phase could be detected. The major reaction into β-SiAlON occurred at 1530°C. The degree of substitution, z, was determined to be 0.6 at 1430°C but increased to 1.0 at 1530°C. β-SiC was detected at all temperatures above 1200°C. The Al was completely converted to Al_2O_3 and AlN, respectively.

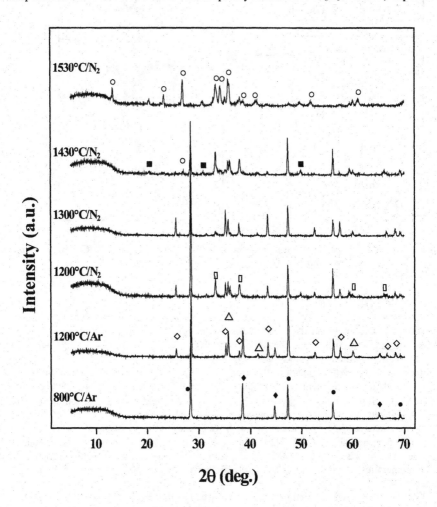

Figure 2: *XRD spectra of the specimens after pyrolysis up to 1200°C for 1 h in Ar and after nitridation at different temperatures for 3 h (O = β-SiAlON, △ = β-SiC, ■ = β-Si_3N_4, ▫ = AlN, ◇ = αAl_2O_3, ◆ = Al, ● = Si).*

Fig. 3 shows SEM-micrographs of samples, pyrolysed at 1200°C, after nitridation at different temperatures. Fig. 3a shows a junction of the corrugated cardboard of a sample after nitridation at 1430°C for 3 h with a cell wall thickness of approximately 200 μm. Fig. 3b shows the microstructure on the surface of a sample after nitridation at 1530°C for 3 h. The surface is covered by hexagonal plate-like grains of β′ SiAlON-phase with a thickness 0.5-2 μm and a diameter of 3-7 μm. The measured skeleton (strut) densities of the SiAlON/SiC composites achieved 2.9 g/cm³ when nitrided at 1430°C and 3.12 g/cm³ for the specimens treated at 1530°C, respectively. While the same geometrical density of 0.44 g/cm³ (corresponding to a porosity of about 85%) was measured for samples nitrided at 1430°C and 1530°C, the skeleton density of the samples after nitridation at 1530°C (3.26 g/cm³) is slightly lower than the theoretical density of β-SiAlON (3.15 g/cm³ for z=1.7) and SiC (3.2 g/cm³) indicating only a low porosity in the strut material of the final SiAlON/SiC-ceramic composites.

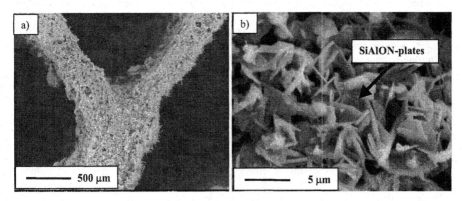

Figure 3: *SEM micrographs of the corrugated cardboard specimens after nitridation for 3h at: a) 1430°C and b) 1530°C.*

CONCLUSIONS

Light-weight, cellular SiAlON/SiC-ceramic composites were produced by Al/Si/polysiloxane infiltration and nitridation of preformed corrugated cardboard structures. Silicon nitride solid solutions with a stoichiometry range of β-$Si_{6-z}Al_zO_zN_{8-z}$ with z = 0.6-1.0 were obtained. The conversion of bioorganic cellulose materials such as paper and corrugated cardboard into ceramic compounds offers new possibilities for manufacturing of highly porous silicon nitride based materials.

ACKNOWLEDGMENTS

The authors thank CNPq-Brazil, the Volkswagen Foundation under contract I / 73 043 for the financial support and Prof. Peter Greil for helpful discussion.

REFERENCES

[1] K.H. Jack and W.I. Wilson, "Ceramics based on Si-Al-O-N and related systems", *Nature (London)*, Phys. Sci., 238 (1976) p. 28.

[2] V.A. Izhevskiy, L.A. genova, J.C. Bressiani and F. Aldinger, "Progress in SiAlON ceramics", *J. Eur. Ceram. Soc.* 20 (2000) p. 2275.

[3] D. R. Lide, *CRC Handbook of Chemistry and Physics, 76th Edition, CRC Press* (1995).

[4] K.H. Jack, "Review: Sialons and related nitrogen ceramics", *J. Mat. Sci.* 11 (1976) p. 1135.

[5] Peelamedu D. Ramesh and Kalya J. Rao, "Preparation and characterization of single-phase β-sialon", *J. Am. Cer. Soc.*, 78 (2) (1995) p. 395.

[6] K. Kishi, S. Umebayashi, E. Tani, K. Shobu and Y. Zhou, " Room temperature strength of β-sialon (z = 0.5) fabricated using fine grain size alumina powder", *J. Eur. Ceram. Soc.* 21 (2001) p. 1269.

[7] I.J. Davies, T. Minemura, N. Mizutani, M. Aizawa and K. Itatani, "Sinterability of β-sialon powder prepared by carbothermal reduction and simultaneous nitridation of ultrafine powder in the Al_2O_3-SiO_2 system", *J. Mat. Sci.* 36 (2001) p. 165.

[8] Q. Li, C. Zhang, K. Komeya, J. Tatami and T. Meguro, "Nano powders of β-sialon carbothermally produced via a sol-gel- process", *J. Mat. Sci. Lett.*, 22 (2003) p. 885.

[9] R.M. Rocha, P. Greil, J.C. Bressiani and A.H.A. Bressiani, "Development and characterization of Si-Al-O-N-C ceramic composites obtained from polysiloxane-filler mixtures", Adv. Powd. Tech. III, Mat. Sci.- Forum, 416 (4) (2003) p. 505.

[10] K. Makuntuala and J.C. Bressiani, "Synthesis of a SiC-SiAlON composite by nitridation of a SiC-AlSi mixture", *Key Eng. Mat.*, 189 (1) (2001) p. 548.

[11] K. Changming, J.-J. Edrees and A. Hendry, "Fabrication and microstructure of sialon-bonded silicon carbide", *J. Eur. Ceram. Soc.* 19 (1999) p. 2165.

[12] P. Greil, "Biomorphous ceramics from lignocellulosics", *J. Eur. Ceram. Soc.* 21 (2001) p. 105.

[13] C.E. Byrne and D.E. Nagle, "Cellulose derived composites - A new method for materials processing", *Mat. Res. Innovat.* 1 (1997) 137.

[14] H. Sieber, D. Schwarze, F. Mueller and P. Greil, "Cellular ceramic composites from preprocessed paper structures", *Ceram. Eng. and Sci. Proc.* 22 (4) 25th Ann. Cocoa Beach Conf. on Composites, Advanced Ceramics, Materials, and Structures: B, ed. by M. Singh and T. Jensen, the Am. Ceram. Soc. (2001) p. 225.

[15] H. Sieber, A. Kaindl, D. Schwarze, J.-P. Werner and P. Greil, "Light-weight cellular ceramics from biologically-derived preforms", *cfi/Ber. DKG*, 77 (2000) p. 21.

[16] M. Scheffler and P. Greil "Polymer derived ceramics. Novel materials with a wide range of potential applications", *Adv. Eng. Mater.* 4 (11) (2002) p. 831.

[17] A.D. Mazzoni, E.E. Aglietti, "Aluminothermic reduction and nitriding of high silica materials (diatomite and bentonite) minerals", *App. Clay Sci.* 17 (2000) p. 127.

[18] C. Zhang, R. Janssen, N. Claussen, "Pressureless sintering of β-sialon with improved green density strength by using metallic Al powder", *Mat. Lett.* 4339 (2003) p. 1.

Thin Films

TITANIUM OXIDE THIN FILM PREPARATION VIA LOW TEMPERATURE SPIN-COATING PROCESS

Weiwei Zhuang and Yoshi Ono
Sharp Labs of America
5700 NW Pacific Rim Blvd.
Camas, WA 98607

ABSTRACT

A stable titanium oxide aqueous liquid has been produced through the hydrolysis of titanium tetrachloride with methacrylic acid. After mixing with acetic acid or 2-methoxy ethanol, we have obtained titanium oxide precursor solutions with variable titanium concentrations. The precursor solutions are used for the preparation of titanium oxide thin films via a low temperature spin-coating process. Titanium oxide thin films were spin-coated onto a PECVD SiO$_2$ layer on a 150mm-silicon (100) substrate. The relationship between the solution concentration and the resulting film thickness has been investigated. For a single layer, a thickness of 260nm film was obtained using a 0.8M concentration precursor solution. By using a 0.4M solution, the film thickness was only 130nm. The film composition was studied by EDX, in which there was no chlorine peak detected. In XRD measurements of the crystalline structure, a strong and clear anatase phase was observed for all the films annealed using rapid thermal anneal (RTA) from 300°C to 700°C. For a lower temperature process, baking at 250°C for 5 minutes, only a weak anatase phase was detected. The film optical properties have been characterized. At a wavelength of 632.8nm, the film exhibited a refractive index of 2.05. The transmittance was better than 90% in the visible light range of 400 to 800nm as measured on a spectrometer.

INTRODUCTION

Because of its unique optical properties and broad range of applications, titanium oxide films have experienced an extensive development over a long period of time. Various deposition technologies have been applied to thin film preparation, including chemical vapor deposition (CVD) [1], sputtering [2], thermal oxidation of titanium [3], sol gel [4, 5] and so on. Recently, there has been a great interest in developing a material with high refractive index, highly transparent to visible light, thickness near 1μm with the ability to be formed at low temperatures. For this purpose, titanium oxide films deposited via a spin-coating sol-gel process is a good choice.

Titanium oxide film depositions via sol-gel processes have been studied over a long period of time [6]. Titanium alkoxides are commonly used titanium sources during the preparation of the spin-coating precursors. Through the hydrolysis by the addition of water and incorporation with special organic solvent, titanium alkoxide precursor solutions led to titanium oxide films via spin-coating process with excellent properties, but the thickness limitation of the film is still an issue. To make thicker titanium oxide films, it is necessary to have some unsaturated organic material involved in the synthesis in order to form a large titanium metalorganic network in solution, or the polymerization of titanium organometallic compound, to reduce the stress produced during the baking process. Thus the titanium oxide nanoparticles (less than 10nm) were prepared and then introduced into polymer solutions to form polymerized titanium organometallic precursor solutions. The use of the solution produced thick titanium oxide films which satisfied optical requirements, but higher baking temperatures are necessary to reduce the carbon impurities in order to increase the film refractive index. Besides, the solution stability is still in question for thick film deposition. In 2003, Rantala et al. [5] reported the deposition of titanium oxide films at low temperature using a titanium tetrachloride based precursor solution, which attracted our attention. After modifying the synthesis procedure and baking process, we have obtained thick titanium oxide films at low temperatures. The films showed excellent optical properties. In this paper, we will present the synthesis, spin coating process and the resulting film properties in detail.

EXPERIMENT

A titanium tetrachloride based precursor solution was synthesized in two steps. In the first step, a concentrated solution was produced, and then in the second step, it was diluted with 2-methoxy ethanol to a specific concentration for thin film deposition via spin-coating process. The synthesis of the concentrated solution was processed using the air-free Schlenk line operation technique. A three-neck round bottle flask was equipped with a graduate cylinder, a rubber stopper and an adapter connected to the Schlenk line. After removing the air and refilling the flask with Ar gas, dichloromethane was transferred into the flask. The calculated volume of $TiCl_4$ was then placed in the graduated cylinder. The $TiCl_4$ liquid was dropped into the flask and mixed with dichloromethane. The methacrylic acid was added into the flask with a syringe through the rubber stopper. The solution was vigorously stirred and the color changed to deep red. In the next step, water was introduced into the solution through the graduated cylinder. During the addition of water, solid precipitates were formed which then dissolved with the addition of more water. After stirring for several hours, the aqueous solution was extracted from dichloromethane, and the titanium concentration in the aqueous solution was above 4M. The solution was in deep red color and stable at room temperature over a long time. To make the solution for titanium oxide deposition via spin-coating process, the concentrated solution was diluted with 2-methoxy ethanol to the targeted concentration.

The titanium precursor solution concentration was determined from the weight of TiO_2 contained in the measured precursor solution volume. The titanium oxide was obtained by slowly heating the solution to 900°C and exposing in air for oxidation for additional two hours.

Titanium oxide films were spin-coated onto 150mm Si wafers with 400nm of PECVD silicon oxide. The standard spin-coating process included the manual dispersion of the precursor solution on the wafer surface, and then spreading the solution over the wafer surface with a 300rpm slow spin speed for 10 seconds, and a 2000rpm high spin speed for 30 seconds. The baking process includes three hot plates installed with auto-transfer between stations. The temperatures of the three hot plates were in the sequence of 100, 180, and 250°C, with the bake time on each stage set to 2 minutes. The temperature on each hot plate is very uniform, with a maximum of 2° deviation.

Film thickness and optical properties were measured by using a Sentech SE-800 Spectroscopic Ellipsometer. ^{1}H NMR spectra were recorded using a Bruker 300MHz NMR instrument. UV-visible spectra were obtained using Jasco V-550 UV/VIS spectrophotometer.

RESULTS AND DISCUSSION

Titanium organometallic precursor solution was synthesized by using titanium tetrachloride as the titanium source, and the unsaturated organic compound methacrylic acid as the additive to improve the thin film deposition stability. The addition of methacrylic acid leads to the reaction between acid and titanium tetrachloride to produce titanium carboxylic compound, as shown in equation (1). The reaction was confirmed from ^{1}H NMR experiment as shown in Figure 1. A hydride peak at 0.51 ppm was observed by the addition of TiCl$_4$ into methacrylic acid CD$_2$Cl$_2$ solution. The hydride peak was assigned to the reaction by-product HCl. The titanium methacrylate ligand can go further condense through its unsaturated double bond to form a large titanium organometallic network, as shown in the equation (2). The polymerization of titanium organometallic compound greatly reduced the stress produced during the titanium oxide thin film baking process by slowing the precursor decomposition rate.

$$xCH_3C(CH_2)COOH \quad + \quad TiCl_4 \quad \rightarrow \quad [CH_3C(CH_2)COO]_xTiCl_{4-x} \quad + \quad xHCl \qquad (1)$$
$$n[CH_3C(CH_2COO\text{-}Ti] \quad \rightarrow \quad \text{-}[\text{-}CH_2(CH_3)C(COOTi)(CH_2)C(COOTi)(CH_3)\text{-}]_n\text{-} \qquad (2)$$

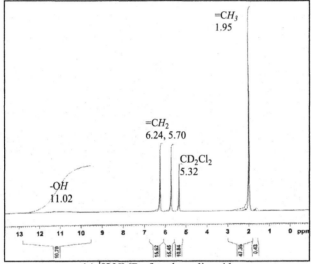

(a) ^{1}H NMR of methacrylic acid

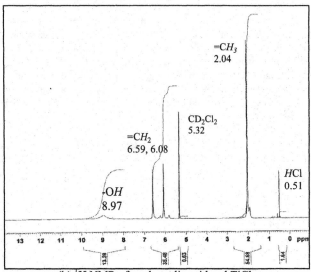

(b) ^{1}H NMR of methacrylic acid and TiCl$_4$

Figure 1. ^{1}H NMR spectra (300 MHz, CD$_2$Cl$_2$)

To achieve a stable and repeatable process, the concentration of the titanium precursor solution must be strictly controlled. It is necessary to establish a standard reference to determine the concentration of all new synthesized precursor solutions. For this purpose, we have prepared a series of standard solutions by mixing one part of concentrated solution with several parts of 2-methoxy ethanol. The solution concentrations were then characterized, and the results were shown in Table 1. Corresponding to the experimental concentration data, we have recorded UV-visible spectra of the series standard solutions, as shown in Figure 2. There is a linear relationship between the solution concentration and the solution transmittance at 475nm. With the standard reference of Figure 2, a repeatable spin-coating process has been approached via the control of every batch of spin-coating precursor concentration and the standard spin-coating process.

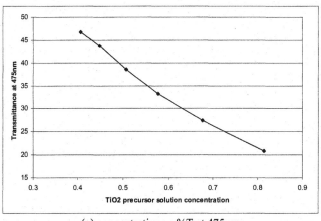

(a) concentration vs %T at 475nm

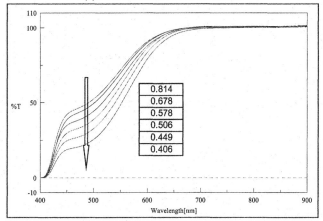

(b) concentrations vs transmittance

Figure 2. Relationship between solution concentration and UV spectra

Table 1. Solution concentration characterization

Items	Ti-10	Ti-14	Ti-15	Ti-16	Ti-17	Ti-18	Ti-19
Ti-solu	1	1	1	1	1	1	1
2-ME	0	4	5	6	7	8	9
conc. (M)	4.2494	0.814	0.678	0.578	0.506	0.449	0.406

The relationship between the solution concentration and the titanium oxide film thickness has also been explored in Figure 3. The films were deposited using our standard spin-coating process. The film thickness was that of a single coating. The results indicated a close linear relationship between the solution concentration and the film thickness. With the increase in the solution concentration from 0.4 to 0.8M, the titanium oxide film thickness increases from 125 to 270nm.

Figure 4 is the spin-rate effect on the titanium oxide film thickness. The increase of spin-rate from 1500 to 3000rpm results in a decrease in film thickness from 220 to 148nm. Our standard spin-coating process was 2000rpm, which gave a thickness of around 180nm.

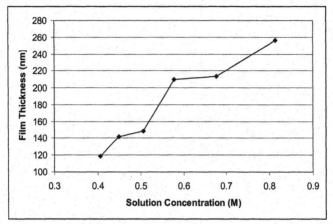

Figure 3. Thickness vs concentration

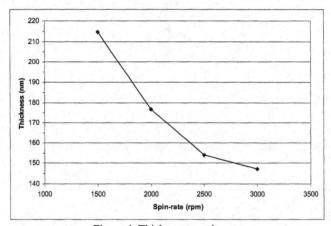

Figure 4. Thickness vs spin-rate

Four layers of titanium oxide have been obtained with the baking sequence modified to 110°C, ≤220°C, and 90°C, each for 2minutes. Optical properties are shown in Figure 5. The increase in baking temperature leads to an increase in the measured refractive index, in which it is higher than 1.9 for the 220°C baked sample. The extinction coefficient is low for all of the samples, and basically, these titanium oxide thin films do not have any absorption of light in the visible region. The increase in baking temperatures also correlates with a decrease in the film thickness, as shown in Figure 6. The four layer titanium oxide film baked at 220°C has a thickness of 573nm, while one baked at 190°C was 617nm. If the baking temperature was increased above 220°C, the four layer titanium oxide film begins to exhibit cracks. Figure 7 shows the optical properties for a 300nm thick film baked at 250°C for two minutes. The refractive index is higher than 2.0 in the 400 to 700nm wavelength range, and has a very low extinction coefficient. The film transmittance spectrum was obtained by spin-coating the film on a quartz plate, the result is shown in Figure 8. Taking the refractive index of the film to be 2, we can calculate the reflected light at the film surface to be 10%, and at the TiO_2/SiO_2 interface to be 2%. Therefore, the result in Figure 8 suggests the film is highly transparent over the entire visible light region.

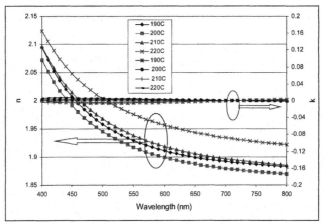

Figure 5. Optical properties vs baking temperature

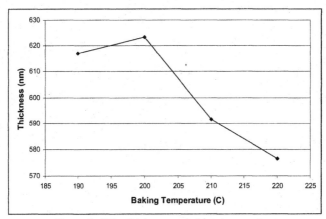

Figure 6. Film thickness vs baking temperature

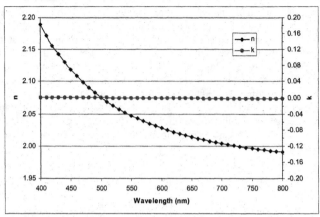

Figure 7. Optical properties for the sample after baking at 250°C for five minutes

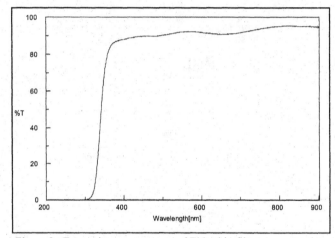

Figure 8. Transmittance spectrum of TiO₂ thin film on quartz plate

The chlorine contamination was analyzed via EDX as shown in Figure 9. Only titanium and oxygen were observed. No chlorine peaks were observed. Thus the titanium oxide films are basically chlorine-free films even though titanium tetrachloride was employed as the titanium source for the precursor synthesis.

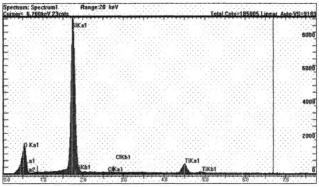

Figure 9. EDX spectrum

The crystallization behaviors of titanium oxide films were studied via X-ray diffraction, as shown in Figure 10 and 11. The increase in annealing temperature resulted in higher crystallization. Only the anatase phase was observed for all of the samples. For the film baked at 250°C for five minutes, a weak anatase phase was detected.

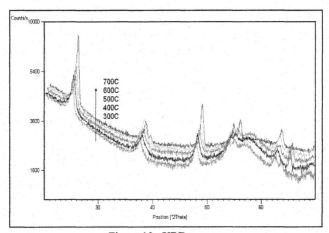

Figure 10. XRD spectra

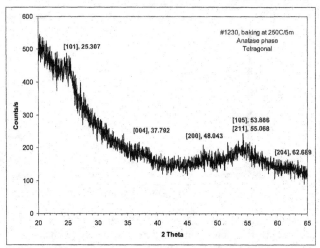

Figure 11. XRD spectrum of TiO$_2$ thin film baked at 250°C for five minutes

SUMMARY

A titanium metalorganic spin-coating solution has been synthesized using titanium tetrachloride as the titanium source. Titanium oxide thin films have been obtained via spin-coating process at low temperature. After modifying the baking process, we have obtained titanium oxide films as thick as 600nm with high refractive index and highly transparent in the visible light region. EDX study indicates a chlorine-free film. The film was crystallized to the anatase phase as indicated by X-ray diffraction measurements.

REFERENCES

[1]L.M. Williams and D. W. Hess, "Structural properties of titanium dioxide films deposited in an rf glow discharge," *J. Vac. Sci. Technol.* **A** [1] 1810-1819 (1983).

[2]B, Bellan, "Dielectric properties of ion plated titanium oxide thin films," *J. Non-cryst. Solids,* **55** 405-412 (1983).

[3]B. Morris Henry, "Method of depositing titanium dioxide (rutile) as a gate dielectric for MIS device fabrication," *US Patent 4,200,474,* (1978).

[4]M. Gratzel, "Sol-gel Processed TiO$_2$ Films for Photovoltaic Applications," *Journal of Sol-gel Science and Technology,* **22** 7-23, (2001).

[5]J.T. Rantala and A.H.O. Karkkainen, "Optical properties of spin-on deposited low temperature titanium oxide thin films," *Optics Express,* **11**[12] 1406-1410 (2003).

[6]K.A. Vorotilov, E.V. Orlora and V.I. Petrovsky, "Sol-gel TiO$_2$ films on silicon substrates," *Thin Solid Films,* **207** 180-184 (1992).

Synthesis of C and N Based Thin Films by Microwave Plasma Chemical Vapor Deposition

R. S. Kukreja, V. Shanov, and Raj. N. Singh
Department of Chemical and Materials Engineering,
University of Cincinnati,
P. O. Box 210012,
Cincinnati, OH 45221-0012

ABSTRACT

Si-C-N based films were deposited on Si (100) substrate using microwave plasma chemical vapor deposition (MPCVD) technique. Effect of variation of CH_4, H_2, N_2, Ar, temperature and pressure on the films obtained was studied. Comparison of the morphology of the films observed under light microscope and SEM with those in the literature suggested the formation of carbon nitride with some Si incorporation into the films. Raman analysis also suggested the formation of carbonitride compound.

KEYWORDS: Microwave plasma CVD, C-N based compounds, thin films, Raman spectroscopy.

INTRODUCTION

In 1989 Liu and Cohen [1] predicted that replacing C for Si in β-Si_3N_4 would result in a high bulk modulus material β-C_3N_4, with hardness comparable to that of diamond. A number of studies were done to synthesize the hypothetical compound and in doing so other compound of C and N [2], like, alpha carbon nitride (α-C_3N_4) analogous to α-Si_3N_4, face centered cubic (c-C_3N_4), pseudo-cubic (Zinc Blende) C_3N_4, and graphitic C_3N_4, were discovered.

Of the various techniques employed for the synthesis of C-N based compounds MPCVD has shown the ability to deposit crystalline C-N based films with high nitrogen incorporation in the crystals. Some of the advantages of MPCVD [3] include high plasma density, easy excitation of species (H^+, N_2^+, C_2^+), wide range of temperature and pressure processing, stability and high energy efficiency, and metastable material (Diamond, c-BN) processing.

In the present paper a preliminary study on deposition of C-N based compounds employing N_2, H_2, CH_4, and Ar gases in the microwave plasma CVD is presented. Effect of pressure and temperature on the films obtained is also investigated. The morphology of the films obtained is studied using optical microscope and scanning electron microscope (SEM). Chemical bonding in the films is analyzed by Raman spectroscopy.

EXPERIMENTAL

Polished Si (100) substrates were used for the deposition of carbon and nitrogen based compounds using MPCVD technique. Scratched and unscratched polished substrates were used to study the effect of surface energy on deposition of C – N based compounds. A base vacuum of 10^{-6} Torr was reached using a combination of turbo-molecular and rotary pump. Gases were fed in the chamber through a gas shower located at the top of the chamber. The flow of the gases was controlled using mass flow controllers. Pressure in the chamber was regulated using a throttle valve. The microwave power was adjusted for reflected power using three screw adapters to achieve zero reflected power. Further information on the MPCVD system used can be found elsewhere [3]. A mixture of semiconductor grade N_2 (99.999%), CH_4 (99.9%), H_2 (99.999%), and Ar (99.9%) gases with different flow rates were used as source gas, typical total flow rate being 100sccm. While the microwave power was controlled at 900W, pressure was varied between 4000 – 12665 Pa, and temperature was maintained between 375 and 750°C, as measured by K-type thermocouple located below the graphite susceptor, using radio frequency (RF) heater. The Si (100) substrates were etched with 5% HF for a few minutes, to remove the native oxide layer. Further etching of the silicon substrate was done using H_2 plasma. Table I. gives the experimental conditions used during deposition.

Table I. Deposition Conditions

Sr. No.	Exp. No.	N_2 sccm	Ar sccm	CH_4 sccm	H_2 Sccm	Power W	Pressure Pa	Temp. °C	Time Hrs
1	122103	50	49	1	E*	900	12665	750	5
2	030204	50	45	5	E*	900	4000	375	5
3	030704	50	45	5	E*	900	4000	750	5
4	032704	50	0	5	45	900	12665	750	5
5	033004	50	0	10	40	900	12665	750	5
6	040104	40	0	10	50	900	12665	750	5

The experimental parameters listed in Table I were decided after careful consideration of the previously preformed experiments by our group [4,5]. Need to add either Ar or H_2 in CH_4 and N_2 plasma was realized. Since at higher temperatures Si catalyses [8] the deposition of C-N based films and high pressure is generally known to increase the deposition rate, effect of both high temperature (750°C) and high pressure (30-95 Torr) were investigated.

Morphology of the films was characterized by a Hitachi S4000 scanning electron microscopy (SEM). The different components in the films were identified by Raman spectroscopy (Instrument SA, T 64000, Jobin Yvon triple monochromator system, equipped with an optical multi-channel detector –CCD array). The output power of the Ar^+ laser (514.5 nm wavelength) was 10 mW, and it was focused up to 2 µm.

RESULTS AND DISCUSSIONS
Morphological Study Using Optical Microscope:
Experiment no. 122103 resulted in Si_3N_4 needles. The needle structure was observed in SEM (Fig. 1.) and was identified as Si_3N_4 using X-ray and Raman analysis techniques. The experiment also caused some melting of Si substrate, indicating additional heating of the substrate by the plasma to close to the melting point of Silicon (1410°C). This suggested that the temperature in the chamber for the experimental conditions is very high and that carbon in the chamber is insufficient. The next experiment, exp. No. 030204 was therefore performed with higher CH_4 in the plasma and at lower temperatures.

E* = Etching only

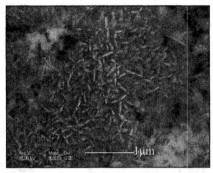

Figure 1. SEM image of Si₃N₄ at 30000X magnification and 10.00 KV Acc. V.

Optical micrograph at 500X magnification of the film deposited using the conditions for experiment no. 030204, as mentioned in Table I, is shown in Figure 2. The spheroids observed in the films show higher density along the scratch lines. The low density of the spheroids was attributed to the low temperature (375°C) used during deposition. During deposition under low temperature conditions the mobility of ionic species on the substrate surface is slow and so is the deposition rate. The next experiment therefore involved higher temperature, maintaining other parameters constant, to see the effect of increase in temperature. Figure 3 shows the optical micrographs of films obtained under conditions of experiment no. 030704 (given in Table I).

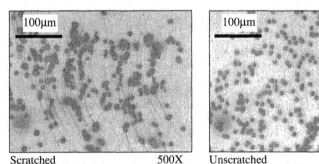

Scratched   500X   Unscratched   500X

Figure 2. Surface morphology of scratched and unscratched specimens at 500X magnification. Scratched specimen shows higher density of spheroids on the scratch lines.

As can be seen from Figure 3 size of the spheroids observed increased with increase in temperature for both rough (unpolished side of Si substrate) and polished/unscratched substrates. However, a decrease in density was observed on the rough surface. This can be attributed to lack of interaction of the plasma species with the substrate material caused by the surface topography.

These results suggest that the surface conditions play a very important role in the deposition of C-N based compounds. Scratch lines on polished substrate surface produce areas of high surface energy and high dislocation density. Both surface energy and dislocation density increase the interaction of the plasma species with the substrate material and thus increase the nucleation density. Rough (unpolished side of Si substrate) surface as observed in Figure 3 although constitutes area of high dislocation density but its topography somehow does not allow the proper interaction of the plasma species with the substrate material resulting in low density of spheroids.

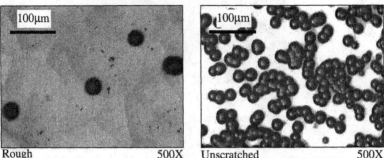

Rough   500X   Unscratched   500X

Figure 3. Surface morphology of rough and unscratched specimens at 500X magnification.

It is well know for deposition of diamond films with MPCVD using Ar/H_2 and CH_4 plasmas that deposition rate increases exponentially with deposition pressure. The reason being that with Ar/H_2 in the plasma increasing pressure increases the efficiency of ionization of CH_4 and thus the super saturation of C_2^+/C^+ radicals, which are the building blocks of diamond thin films. However, it is also known that incorporation of N in $Ar + CH_4$ plasma produces ultrananocrystalline diamond [10] and not carbon nitride. Therefore, in line with the above explanation experiment no. 032704 was performed with two changes, one, the pressure was increased to 95 Torr, and second, Ar was replaced with H_2.

Figure 4. shows the optical micrograph of films obtained, on scratched and polished Si (100) substrates, from experiment no. 032704. It is seen that with H_2 in the plasma, the density of spheroids has increased drastically on the scratched substrate. While on the polished (unscratched) substrate along with spheroids some rod-like crystals (RLCs) are also observed.

Silicon is know to catalyze the deposition of crystalline C and N based compounds [5]. At higher temperatures (>700°C) the diffusion of Si in the depositing C-N based film increases. The increase in the Si content in the film increases the chance of formation of crystalline Si-C-N film. Radicals such as C^+ formed from dissociation of CH_4 in the plasma constituting $CH_4 + H_2 + N_2$ combine with the N^+ ions in the plasma forming CN radicals, which are the building block for C-N based compounds. RLCs have been observed by others and suggest as Si-C-N compound with sp^3 coordination of atoms.

The observation of the chamber conditions after deposition of experiment no. 032704 did not show any soot formation or carbon deposition on the walls. This indicated that most of the carbon that was introduced into the chamber as CH_4 gas is being used either in forming spheroids and RLCs, or other non depositing species like HCN, CH_3, and CH, which recombine and exit as a mixture of gases. This information also tells us that more CH_4 can be added to the plasma without causing deposition of carbon soot. The above interpretation was supported by our next experiment, experiment no. 303004, which included increase in CH_4 flow rate. Since the total flow rate was maintained at 100sccm, hydrogen flow rate was reduced to accommodate the increase in CH_4 flow rate.

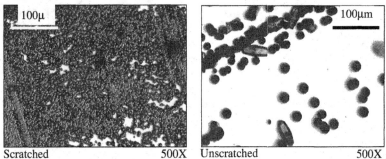

Scratched   500X   Unscratched   500X

Figure 4. Surface morphology of scratched and unscratched specimens, of exp. no. 032704, at 500X magnification. Unscratched specimen shows higher density of spheroids while on the polished surface rod like crystals (RLCs) along with spheroids are observed.

With increase in CH_4 in the plasma an increase in the density of RLCs and spheroids was observed on the unscratched / polished silicon substrate, while a very high density of spheroids was observed on the scratched surface (Figure 5). This indicates that both RLCs and spheroids are C based compounds. Also with a decrease in H_2 in the plasma more of C^+ radical is available for the formation of CN film. Therefore it may be suggested that increased density of spheroids may be due to the decreased H_2 in the plasma. To confirm the correlation between H_2 in the plasma and the density of spheroids next experiment was performed with increased H_2 in the plasma. The increase in H_2 was accommodated with decrease in N_2 in the plasma gas, thus maintaining the total flow of 100sccm. Figure 6. shows the optical micrographs of films deposited using the deposition conditions of experiment no. 040104.

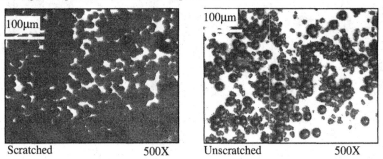

Scratched   500X   Unscratched   500X

Figure 5. Surface morphology of scratched and unscratched specimens, of exp. no. 033004, at 500X magnification. Unscratched specimen shows higher density of spheroids while on the polished surface rod like crystals (RLCs) along with spheroids are observed.

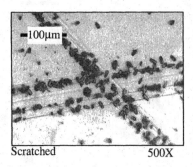 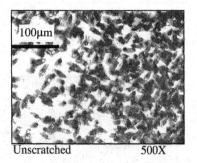

| Scratched | 500X | Unscratched | 500X |

Figure 6. Surface morphology of scratched and unscratched specimens, of exp. no. 040104, at 500X magnification. Only RLCs and no spheroids are observed on both the substrates. Unscratched specimen shows preferential deposition of RLCs on the scratch lines.

As shown in Fig. 6 the density of RLCs increased on both scratched and unscratched substrates. Spheroids are almost absent in this case. Thus, it is reconfirmed that the density of spheroids depend on the concentration of H_2 in the plasma. Scratched specimen shows preference for the RLCs to grow on the scratch lines, which are areas of high dislocation density.

Morphological Study Using SEM

Figure 7 shows the SEM images of RLCs (032704) and Spheroids (030204). It is observed that RLCs are well faceted hexagonal crystals of 1 – 4 μm and show both primary and secondary nucleation. Spheroids appear to be agglomerates, of 1-2 μm, of many smaller grains.

 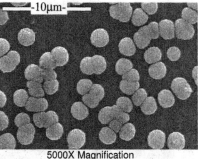

| 5000X Magnification | 5000X Magnification |

Figure 7. Surface morphology of scratched and unscratched specimens, of exp. No. 032704 and exp. no. 030204, at 5000X magnification.

Comparison of Results With Literature:

Some authors have shown similar morphological results using MPCVD and hot filament CVD (HFCVD). A comparison of our results with those observed by others in the literature suggests that our films are Si-C-N based compounds. Table II shows a comparison between our results and those obtained in the literature.

Table II. Comparison of experimental research with literature.

Ref. No.	Tech.	Sub.	CH4 (sccm)	N2 (sccm)	H2 (sccm)	P (W)	Pr (Torr)	Temp. (°C)	Film
	MPCVD	Si	10	40	50	900	95	750	Si-C-N
6	MPCVD	Si	0.5–1.0	100-80	0	500-700	--	800–950	C3N4
7	MPCVD	Si, HOPG	0.7	100	0	--	16 - 20	810	G-C3N4
8	MPCVD	Si	20	80	80	1500	30	400-1200	Si-C-N
9	MPCVD	Si	1.0 - 10	100	0	150-200	25	1100 - 1200	A-C3N4 / Si3N4
10	HFCVD	Ni, Si	0.5	70	0	--	0.5 - 15	750 – 950	C-N/ Si-C-N
11	MPCVD	Si	1 – 4	0 – 98	0 - 198	1000	30	800-900	Si-C-N

Raman Analysis

Figure 8. shows Raman peaks for spheroids (030704) and RLCs (040104). It is seen that similar peaks are observed for both morphologies. Increased intensity of Raman peaks for RLCs can be associated with highly ordered structure of RLCs, as seen from SEM, as compared with spheroids. The sharp peak with very high intensity at 522.4 cm-1 is of Si(100) substrate.

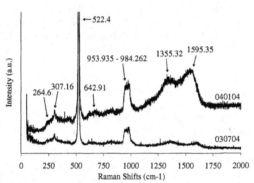

Figure 8. Raman spectra of scratched samples of 040104 and 030704 shows same peaks for both spheroids (030704) and RLCs (040104).

Table III lists calculated Raman peaks for β-C_3N_4, observed Raman peaks for β-Si_3N_4 and Raman peaks observed in the deposits obtained in our experiments. It is observed that the peaks observed for deposits in our experiments match closely with those theoretically calculated for β-C_3N_4. This suggests that both spheroids and RLCs observed on Si (100) substrate in our experiments are C-N based compounds.

Table III. Calculated Raman peaks for β-C_3N_4, observed Raman peaks of Si_3N_4 [12,13] and observed Raman peaks of the films deposited.

β-C_3N_4 (cm^{-1})	β-Si_3N_4 (cm^{-1})	Experiment numbers					
		040104-1 (cm^{-1})	040104-2 (cm^{-1})	030704-1 (cm^{-1})	030704-2 (cm^{-1})	032704-1 (cm^{-1})	032704-2 (cm^{-1})
206	144	--	--	--	--	--	--
266	186	262	264	255	--	261	--
300	210	310	307.1	303.9	307.8	--	--
327	229	--	--	--	395.8	366	--
645	451	--	642	626.8	630	--	--
672	--	686.4	--	685.8	683.7	682.7	687
885	619	--	--	833.5	838.9	958.6	--
1048	732	982.4	984	980.7	988.4	1052	--
1237	865	--	--	--	--	--	--
1327	928	--	--	--	--	--	1334
1343	939	1380.7	1355	--	1348	--	--
1497	1047	1582.8	1595	--	1548	--	1564

We consider the work presented here as preliminary research and a lot of further investigations are under way. Some of these include using XPS, TEM, FTIR and RBS in order to understand more precisely the composition and deposition of the deposit. Experiments with longer deposition time and parameters suitable for deposition of only RLCs and only spheroids to produce uniform and thick films are planned. Effect of different substrate materials like Ni, a-Si_3N_4, and diamond films will be explored in the near future. Influence of Boron addition in the plasma gas on the deposit and formation of B-C-N compounds is also under investigation.

CONCLUSIONS:
Our conclusions are based on the preliminary results reported in this paper. MPCVD was successfully used to synthesize C and N based compounds on Si(100) substrates. Two major morphologies were observed depending on the conditions, i.e. rod-like crystals and spheroids. High temperature and high pressure plasma promoted the deposition of RLCs. Adding Ar in CH_4 and N_2 plasma resulted in spheroids. Density of spheroids in CH_4+N_2+H_2 plasma depended on concentration of H_2 in the source gas. Higher the concentration of H_2 in the source gas lower was the density of spheroids. Comparison of these results with those in the literature and Raman analysis showed that both the spheroids and the RLCs were C-N based compounds.

ACKNOWLEDGEMENTS
The authors would like to thank Dr. Punit Boolchand and Dr. Sergey Mamedov for the Raman Spectroscopy measurements, and Mr. Srinivas Subramanian for the SEM images. This material is based on work supported by the Nations Science Foundation under grant No: CMS-0210351. Any opinions, findings, and conclusions or recommendations expressed in this material are those of the author and do not necessarily reflect views of the Nations Science Foundation.

REFERENCES:

[1] A.Y. Liu and M.L. Cohen, "Prediction of new low compressibility solids," *Science*, **245** 841 (1989).

[2] Z. J. Zhang, J. Huang, S. Fan and C. M. Lieber, "Phases and physical properties of carbon nitride thin films prepared by pulsed laser deposition," *Materials Science and Engineering*, A **209** 5 – 9 (1996).

[3] V. Jayaselam, "Processing of polycrystalline diamond films by microwave plasma CVD", MS thesis (2000).

[4] R. Ramamurti, R.S. Kukreja, Li Guo, V. Shanov and R.N. Singh, "Microwave plasma chemical vapor deposition (CVD) of carbon based films in the system C-N," Presented at the 28th International Conference and Exposition on Advanced Ceramics and Composites, Focused Session C: Nanomaterials and Biomimetics, January 25th – 30th, 2004, Cocoa Beach, Florida.

[5] R. Ramamurti, V. Shanov and R.N. Singh, "Synthesis of nanocyrstalline diamond films by Microwave plasma CVD," *Proceedings of American Ceramic Society, Session: Innovative Processing/Synthesis: Ceramics, Glasses, Composites* VI, 39 – 50 (2002).

[6] Y. Zhang, H. Gao and Y. Gu, "Strucuture studies of C_3N_4 thin films prepared by microwave plasma chemical vapor deposition," *J. Phys. D: Appl. Phys.*, **34** 299-302 (2001).

[7] L.P.Ma, Y.S. Gu, Z.J. Duan, L. Yuan and S.J. Pang, "Scanning tunneling microscopy investigation of carbon nitride thin films grown by microwave plasma chemical vapor deposition," *Thin Solid Films*, **349** 10 –13 (1999).

[8] L.C. Chen, C.K. Chen, S.L. Wei, D.M. Bhusari, K.H. Che, Y.F. Chen, Y.C. Jong and Y.S. Huang, "Crystalline silicon carbon nitride: A wide band gap semiconductor," *Applied Physics Letters*, vol. **72**, No. 19, 2643 – 2465, May 1998.

[9] Y. Sakamoto and M. Takaya, "Fabrication of nitrogen included carbon materials using microwave plasma CVD," *Surface and Coating Technology*, **160 – 170** 321 – 323 (2003).

[10] D. J. Johnson, Y. Chen, Y. He and R.H. Prince, "Deposition of carbon nitride via hot filament assisted CVD and pulsed laser deposition," *Diamond and Related Materials*, **6** 1799 – 1805 (1997).

[11] Y. Fu, C. Q. Sun, H. Du and B. Yan, "From diamond to crystalline silicon carbonitride: effect of introduction of nitrogen in CH_4/H_2 gas mixture using MPCVD," *Surface and Coating Technology* **160** 165 – 172 (2002).

[12] Yen, Tyan-Ywan; Chou, Chang-Pin, "Growth and characterization of carbon nitride thin films prepared by arc-plasma jet chemical vapor deposition," *Applied Physics* Letters, **67** (19), 6 Nov 1995

[13] Y.P. Zhang, Y.S. Gu, X.R. Chang, Z.Z. Tian, D.X. Xhi and X.F. Zhang, "On the structure and composition of crystalline carbon nitride films synthesized by microwave plasma chemical vapor deposition," *Material Science and Engineering*, **B78** (2000) 11 –15.

EFFECT OF NITROGEN ON MORPHOLOGY AND ELECTRICAL PROPERTIES OF POLYCRYSTALLINE DIAMOND THIN FILMS

R. Ramamurti, V. Shanov, R. N. Singh
Department of Chemical and Materials Engineering,
University of Cincinnati,
P. O. Box 210012,
Cincinnati, OH 45221-0012

ABSTRACT

Polycrystalline diamond (PCD) was successfully deposited on silicon (100) with methane precursor in nitrogen/ hydrogen. Nitrogen was incorporated to obtain an n-type diamond films. The effect of varying nitrogen concentrations on the grain size and the quality of the diamond films was investigated. The films were characterized for electrical conductivity, thickness, grain size, and preferred orientation. These results will be presented and discussed.

Keywords: microwave plasma CVD, nitrogen plasma, doped-diamond films, electrical properties

INTRODUCTION

It is well known that the outstanding properties of diamond synthesized by Chemical Vapor Deposition (CVD) diamond such as extreme hardness, wide band gap, transmittance from ultraviolet to the infra-red range, high electrical resistivity and extremely good thermal conductivity make it useful in a wide range of applications. Currently, large efforts are made to utilize these properties in mechanical, optical, electronic and thermal applications.

Considerable amount of work has been performed in order to study the nature of impurities in diamond thin films. The role of grain boundaries in determining the properties of polycrystalline diamond films is especially important. Even heteroepitaxial diamond has a significant amount of defects, leading to sp^2-bonded carbon at the grain boundaries [1-3]. Such grain boundaries are supposed to alter the properties of CVD diamond resulting in improved field emission characteristics, increased electronic conductivity with reduced carrier mobilities, and a decrease in the film hardness [4]. Impurities of different atomic species in diamond have also generated a significant amount of interest, primarily as dopants for electronic devices. One of the best known impurities in both the natural and synthetic diamond is nitrogen [5]. Most of the current work is on finding a good dopant for n-type diamond. This is difficult since nitrogen has a deep donor level in diamond of 1.7 eV below the conduction band. Other n-type dopants like sulfur and phosphorous have been attempted, but only with limited success [6, 7]. Phosphorous has a donor level of 0.55 eV, which could be a better dopant than nitrogen [8]. However there might be some passivation of the dopant atoms by defects or hydrogen atoms in the CVD diamond film, thus reducing electrical activation [8].

Nitrogen doping of ultrananocrystalline diamond (UNCD), in particular is of considerable interest for a variety of reasons. Recently, it has been reported that the conductivity of UNCD increases as much as five orders of magnitude to 140 $(ohm. cm)^{-1}$ when nitrogen is added to the plasma during growth [9]. Theoretical models predict that UNCD grain boundary defects produce a number of states in the band gap of diamond; π and π^* states for the sp^2-bonded atoms, and σ and σ^* states for the sp^3-bonded carbon atoms [10]. When enough of these states form a continuous band of delocalized energy states, across which hopping conduction proceeds [11].

Small quantities of nitrogen in the plasma improve the morphology of diamond films [5, 12]. However, increasing the quantity of nitrogen can have deleterious effects on the deposited films [12]. Microwave plasma CVD studies have reported that the diamond morphology and growth rate depend strongly on the C/N ratio in the input gas as well on the microwave power [13]. Moustakas et. al. [14] found that the growth rates were little affected by the addition of nitrogen upto 1%, but at higher concentrations they observed higher growth rates and a change in morphology from (111) to predominantly (100) facets. The substitutional incorporation of nitrogen results in significant distortion of the diamond lattice. This is because nitrogen being three-fold coordinated, when introduced into the diamond lattice causes formation of a lone-pair of electrons. Early workers attributed this distortion to Jahn-Teller effect [15] and predicted that the C-N bond along the (111) direction should be 10-14% longer than the C-C bond [14]. The deposition of textured diamond has been made possible by incorporation of nitrogen in the diamond films [5]. The thermodynamic calculations [13] and mass spectrometry results show that HCN is the predominant N-containing gas under deposition conditions. The study of the effect of nitrogen on Optical Emission Spectra (OES) has been observed by Vandevelde et. al. by looking at the competing growth mechanisms of CN and C_2 with increasing nitrogen concentrations in the plasma [16]. At high nitrogen concentrations, the carbon super-saturation on the substrate is reduced due to abstraction of adsorbed hydrogen atoms caused by CN and HCN [17]. Since the presence of nitrogen is theoretically predicted to increase the amount of π-bonded carbon in UNCD [9], it becomes very important to accurately characterize the ratio of sp^3-bonded to sp^2-bonded carbon in UNCD. Raman spectroscopy is a very good tool to perform such an analysis. This technique can detect small amounts of non-diamond carbon in the diamond films. Some of the challenges include differences in the scattering cross-section of diamond and graphite [18]. Second, the Raman scattering is dependent on the long-range order of the material, and has a strong dependence on the diamond crystallite size [19].

The growth of nitrogen-doped diamond films by microwave plasma CVD has been investigated in this paper. The characterization for thickness, grain size, preferred orientation and electrical conductivity of the films is also studied. This will be correlated with changes in dominant growth species in the plasma. The nitrogen dopant and carrier concentration in the films will also be determined. It will be verified whether the high levels of nitrogen dopants in the diamond films causes the disruption of the diamond lattice or the gas phase chemistry is not suitable for formation of diamond films with high concentrations of nitrogen in the plasma.

EXPERIMENT DETAILS

An Electron Cyclotron Resonance Microwave Plasma Chemical Vapor Deposition (ECR-MPCVD) facility was used for advanced thin film deposition, including synthesis of the polycrystalline diamond. This facility was based on an ASTEX magnetized microwave plasma source and details about it can be found in our previous publications [20, 21]. A schematic of the CVD system is shown in Fig. 1. Silicon wafers, p-type (100) diced into squares (25 x 25 x 0.5 mm) was used as substrates for deposition of PCD. The silicon wafer has a native oxide layer on it. Silicon with an oxide layer is hydrophilic and bare silicon is hydrophobic. In order to remove this initially a water break test is done to verify if the surfaces are hydrophilic. A dilute HF solution (2.5%) with 25 ml 49% HF and 475 ml deionized water is prepared in polypropylene beakers. The wafer is soaked in this dilute HF solution for a few minutes and then rinsed in water several times. The water break test is done again to see if the surfaces are hydrophobic. The substrate activation procedure included ultrasonic (US) treatment with slurry of 20-40 μm diamond grit for two hours. Prior to deposition, a 30 minute etching was done in the hydrogen plasma to clean the Si substrate off the native oxide layer. Mixtures of CH_4, N_2 and H_2 were used as the reactant gases for the microwave discharges. An optimized set of growth parameters included; substrate temperature of 750 °C, pressure of 30 torr, microwave power of 900 W. The flow rate of CH_4 was kept constant at 1 sccm, while the flow rate of N_2 was varied from 0.1% to 10% and supplemented by H_2. The deposition time was maintained for 8 hours. The films were characterized for thickness, grain size using a Hitachi S4000 scanning electron microscope (SEM). The different components in the film were identified by Raman spectroscopy using a Micro-Raman (Instrument SA, T 64000, Jobin Yvon triple monochromator system, equipped with an optical multi-channel detector-CCD array). The output power of the Ar^+ laser (514.5 nm wavelength) was 10 mW, and it was focused up to 2 µm diameter (objective x80). The X-ray diffraction studies were done using a Rigaku D-2000 powder diffractometer supported by software to analyze Bragg line shapes and phase analysis with JCPDS on file. An MKS Close-Ion source Quadrupole Mass Spectrometer (QMS) was used to identify in-situ the characteristic ionic and molecular species present in the plasma.

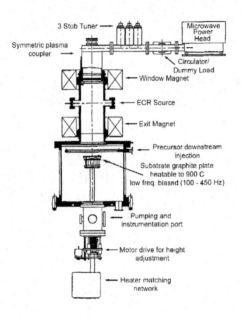

Fig. 1. Schematic of the ECR-Microwave Plasma CVD, based on a magnetized plasma source [22].

RESULTS AND DISCUSSIONS

A study of increasing concentration of nitrogen in the hydrogen and hydrogen/ argon plasma was done to investigate the effect on morphology and grain size of the films. The SEM pictures of nitrogen incorporated diamond films are shown in Fig. 2. It was observed that the grain size decreased from 250 nm to < 10 nm on varying nitrogen from 0.1% to 10%. It is also observed that there is a change from the regular faceted morphology to more irregularly shaped grains. The grain size decreases with increasing nitrogen upto 10%. Clustering of nanocrystalline grains begins at 5% N_2. There is drastic reduction of grain size to the nanometer range beyond 5% N_2. This is also seen in our previous results [23]. This might be due to the distortion from three-fold coordinated nitrogen going into the four-fold coordinated diamond sites and hence leaving a lone-pair orbital. This is attributed to the Jahn-Teller effect [15] and prediction that the C-N bond along the (111) direction should be 10-14% longer than the C-C bond to preserve (111) axial symmetry [14].

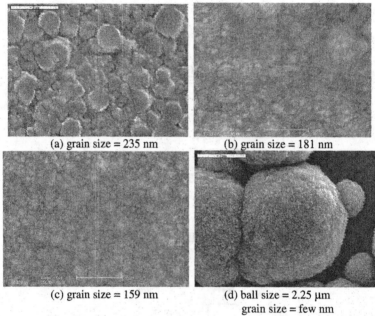

(a) grain size = 235 nm   (b) grain size = 181 nm

(c) grain size = 159 nm   (d) ball size = 2.25 μm
                             grain size = few nm

Fig. 2. Plane-view SEM images of diamond films grown under conditions of 750 °C, 30 torr, 900 W, 1% CH_4, rest H_2 and (a) 0.1% N_2; (b) 0.5% N_2; (c) 1% N_2; (d) 10% N_2

The cross-section SEM images for the same diamond films are shown in Fig. 3a, 3b, and 3c. There is a change from smooth and continuous grain-like structure at 0.1% N_2 to columnar structure at 0.5% and 1% N_2. The growth rate change with nitrogen in the plasma is shown in Fig. 3d. There is no significant change in the growth rate below 1% nitrogen, whereas beyond 1% N_2, the growth rate decreases drastically.

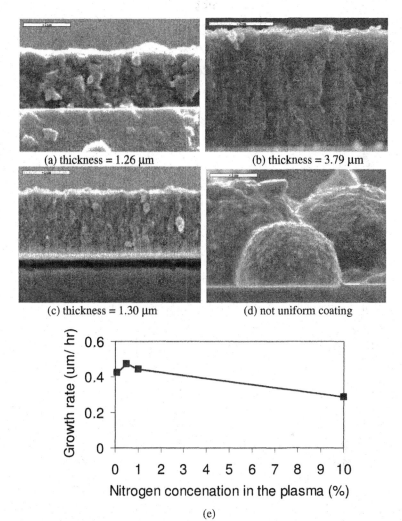

(a) thickness = 1.26 μm (b) thickness = 3.79 μm

(c) thickness = 1.30 μm (d) not uniform coating

(e)

Fig. 3. Cross-sectional SEM images of diamond films grown with under conditions 750 °C, 30 torr, 900 W, 1% CH_4, rest H_2 with (a) 0.1% N_2; (b) 0.5% N_2; (c) 1% N_2, (d) 28% N_2, (e) Growth rate variation of the nitrogen incorporated diamond films with increasing nitrogen in the plasma.

The Raman spectra for the above experiments are shown in Fig. 4. There is a variation in the 1332 cm^{-1} diamond peak intensity and width. Nanocrystalline diamond peak appears at 1150 cm^{-1}, which increases with N_2 and is more prominent at 10% N_2. This indicates that the grain size is decreasing to the nanometer range. The graphite "G" peak at 1550 cm^{-1} decreases in intensity too, but its width decreases with increasing N_2. This means that the graphite content in the film is increasing with increasing nitrogen.

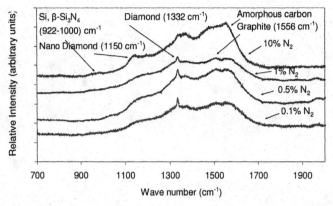

Fig. 4. Raman spectra of diamond films grown with 0.1% - 10% N_2 concentrations in the plasma.

The variation of intensity and width of the diamond peak (1332 cm^{-1}) with the nitrogen content in the plasma is revealed in Figs. 5a and 5b. The diamond peak intensity decreases with a maximum at 0.1% N_2 as seen in Fig. 5a. The width of the peak decreases gradually from 0.1% to 0.5% N_2 in the plasma and then increases beyond that to 10%. These results suggest that there is a decrease in the quality of the diamond films at higher nitrogen contents.

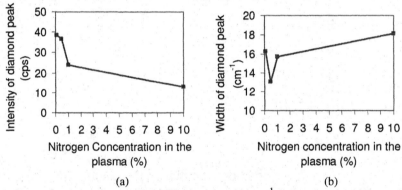

(a)   (b)

Fig. 5. Change in (a) Amplitude and (b) Width of the 1332 cm^{-1} diamond peak with increasing nitrogen concentrations in the plasma.

A quantitative analysis of the Raman spectra was done in order to determine sp^3/ sp^2 ratios in the diamond films. This was done by assigning certain peaks to quantify diamond and some to quantify graphite. Peak Fit software [24] used to perform the fitting of these peaks. Table I shows the results from this analysis. Since the scattering factor of graphite is 57 times that of diamond when using an Ar$^+$ laser of wave length 514.5 nm, we have to take this into account in our calculations. The following formulae were used to determine the sp^3/ sp^2 ratios.

$$R_{diamond} = A_{diamond}/ (A_{diamond} + (1/57)*A_{graphite}) \qquad (1)$$

$$R_{graphite} = (1/57)*A_{graphite}/ (A_{diamond} + (1/57)*A_{graphite}) \qquad (2)$$

Where,

$A_{diamond}$ = area of diamond peaks
$A_{graphite}$ = area of graphite peaks

Table I. Diamond/ Graphite contents in the films with varying nitrogen concentrations in the plasma.

S. No.	Nitrogen Concentration in the plasma (%)	Diamond Content (%)	Graphite Content (%)
1	0.1	88.04	11.96
2	0.5	91.97	8.03
3	1	96.59	3.41
4	10	71.52	28.48

The quantitative analysis shows a maximum in the sp^3 content at 1% nitrogen and there is a drastic decrease at 10% nitrogen. The decrease in grain size with increasing nitrogen indicates that there is a probable increase in the sp^2 phase as confirmed by the Raman study. However, further studies using Electron Energy Loss Spectroscopy (EELS) is needed to confirm that this sp^2 phase is indeed in the grain boundaries.

The X-ray diffraction studies show that there is a random orientation of the diamond grains as seen in Fig. 6a. The D (111) peak is strongest (100 % intensity) for polycrystalline diamond. D (111)/ D (220) Intensity was plotted with increasing nitrogen concentrations in the plasma in Fig. 6b. This was found to decrease, except for a maximum at 0.5% N_2. This indicates a decrease in D (111) texture of diamond films beyond 0.5% N_2.

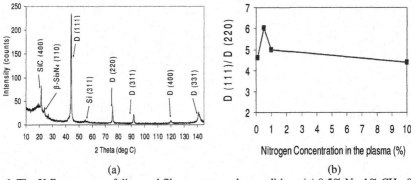

(a) (b)

Fig. 6. The X-Ray spectra of diamond films grown under conditions (a) 0.5% N_2, 1% CH_4, 99% H_2, 750 °C, 900 W, 8 hours and (b) Variation of D (111)/ D (220) intensity with different N_2 concentrations in the plasma.

The QMS study showed that the species in the plasma contributing to diamond growth change with increasing nitrogen concentrations in the plasma. The plasma species monitored during this process are H_2^+, N_2^+, N^+, CH_4^+, CN^+, and HCN^+. These results are given in Fig. 7. It is observed that below 0.5% nitrogen in the plasma, the partial pressure of N^+ and N_2^+ are much lower than that of CH_4^+, CN^+, and HCN^+. At 1% N_2, all species reach a maximum, but partial pressure of CH_4^+ is much greater than that of CN^+, and HCN^+. However, at 10% N_2, it is just reversed with CN^+, and HCN^+ dominating due to a drop in CH_4^+ partial pressure. This might be one of the reasons why the growth rate and quality of these films are lower. It could also give an explanation for the irregularly shaped grains caused by the distortion of the diamond lattice.

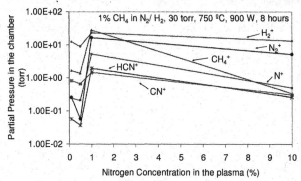

Fig. 7. QMS spectra for monitoring different species with increasing N_2 concentrations in the plasma.

The electrical properties of PCD films were measured by creating test capacitors by (a) high-rate etching of the crystalline host silicon in sulfur hexafluoride (SF_6) plasma, and (b) developing a non-lithographic technique for patterning the silicon to produce free-standing PCD films on which circular (abbreviated: DOT) test capacitors were then fabricated [25]. The free-standing PCD areas were produced by etching away the host silicon over window areas of 5x5 mm^2. A simple method adopted using a shadow mask, which consists of long metal stripes separated by spaces of equal width. Aluminum which was found to be resistant to SF_6 plasma is deposited through the shadow mask over the silicon surface. This produced an array of square windows through which the silicon can be etched away in the SF_6 plasma. After etching, the PCD structure had the appearance of arrays of translucent PCD squares supported on a rigid silicon grid, which provided physical strength to the structures. The high strength and high thermal stability of the PCD allows the sensors to operate at elevated temperatures. Research on such sensors is currently in progress. The ground plane of the capacitor array was fabricated by sputtering a 0.4 mm thick continuous Ti film on the rough PCD surface, and sequentially thermally evaporating 0.1 mm thick gold layer on top of Ti. This is primarily for high temperature testing, but for room temperature measurements Al was found to be a good contact metal. The DOTS on the opposite side of the capacitors were formed using the same metallization procedure through a shadow mask. The DOT PCD structures thus created are shown in Fig. 8.

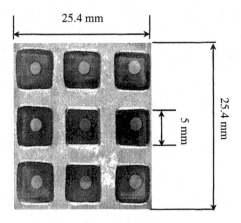

Fig. 8. Typical capacitor test structure of CVD diamond

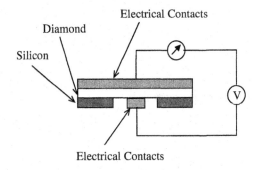

Fig. 9. I-V measurement on free-standing CVD diamond

The I-V measurement was made on this fabricated free-standing CVD diamond as shown in the setup in Fig. 9. Preliminary results for an un-doped CVD diamond were described earlier [25]. For the present work voltage was varied between 10-100 volts and the current measured was in the order of $10^{-4} – 10^{-10}$ amps. The resistivity was measured using the thickness of the film and area of the contact electrodes. This is given by

$$\rho = (R*A)/ t \qquad\qquad (3)$$

Where,

ρ = resistivity of the CVD diamond film (ohm. cm)
A = contact area of the electrode (cm^2)
t = thickness of the film (cm)

The mobility of electrons in CVD diamond for a n-type material is assumed to be $\mu = 1400$ cm^2/volt-sec. The carrier concentration of nitrogen-doped diamond films is determined from the equation [26]

$$n = N_d = \sigma/(e*\mu) \qquad (4)$$

where, σ = conductivity (ohm. cm)$^{-1}$, μ = mobility (cm^2/volt-sec), n = carrier concentration (cm^{-3}), N_d = dopant concentration (cm^{-3})

The carrier concentration is equal to the number of dopant atoms. The conductivity and carrier concentrations were found to increase by 7 orders of magnitude when the nitrogen concentration in the plasma is increased from 0.1% to 10%. These results are shown in Figs. 10a and 10b.

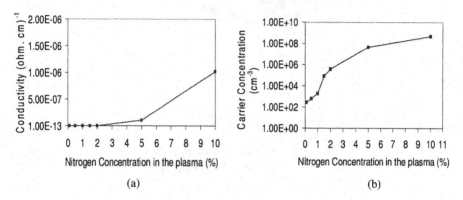

(a)   (b)

Fig. 10. (a) Conductivity and (b) Carrier Concentration variation in the diamond films with increasing nitrogen in the plasma.

These values are lower than that for highly doped (40% N$_2$) diamond which is generally in the order of 10^{19} cm^{-3} [13]. This might be due to the deep donor level of nitrogen (1.7 eV), which may cause low doping efficiency compared to other elements like phosphorous, which has a donor level of 0.55 eV. There might also be some passivation of the nitrogen dopant atom by the defects present in the diamond film.

CONCLUSIONS
Microwave plasma CVD technique explored to grow nitrogen-doped diamond films. The grain size of the films decreased to the nanometer range with increasing nitrogen. Distortion in the diamond lattice was evident from the SEM beyond 1% thus confirming some theoretical predictions. Qualitative Raman analysis of diamond films showed a decrease in quality of diamond with nitrogen. The quantitative analysis of diamond films showed that there was a maximum in the sp^3 content at 1% N$_2$. Decrease in diamond films was also confirmed by X-ray diffraction. Plasma species such as CN$^+$ and HCN$^+$ were found to be dominant at high (> 5%) nitrogen in the plasma. Electrical measurements showed that the conductivity and hence the carrier concentration increased by 7 orders of magnitude with increasing nitrogen in the plasma.

However, the carrier concentration was not very high due to the low doping efficiency of nitrogen. It has to be confirmed whether this was due to any grain boundary phase from EELS. SIMS analysis also needs to be done to determine the variation in the nitrogen concentration in the films with depth. Hall measurements are planned to determine the mobility of the majority carriers in the diamond films. These studies are under way and will be published in the future.

ACKNOWLEDGEMENTS

The authors would like to thank Dr. Punit Boolchand and Dr. Sergey Mamedov for help with the Raman Spectroscopy. The help rendered by Mr. Srinivas Subramanian for obtaining the SEM images is greatly appreciated. We would also like to thank Mr. Mahmood Samiee for helping us with the electrical measurements. This material is based upon work supported by the National Science Foundation under Grant No. DMR-0200839. Any opinions, findings, and conclusions or recommendations expressed in this material are those of the authors and do not necessarily reflect the views of the National Science Foundation.

REFERENCES

[1]M. Schreck, F. Hormann, H. Roll, J. K. N. Linder, and B. Stikzer, "Diamond Nucleation on Iridium Buffer Layers and Subsequent Textured Growth: A Route for Realization of Single Crystal Diamond Films", Appl. Phys. Lett. 78, 192 (2001)

[2]J. B. Posthill, D. P. Malta, T. P. Humphreys, G. C. Hudson, R. E. Thomas, R. A. Rudder, and R. J. Markunas, "Method of Fabricating Free-Standing Diamond Single Crystal using Growth from the Vapor Phase", J. Appl. Phys. 79, 2722 (1996).

[3]J. Isberg, J. Hammersberg, E. Johannson, T. Wilkstrom, D. J. Twitchen, A. J. Whitehead, S. E. Coe, and G. A. Scarsbrook, "High Carrier Mobility in Single-Crystal Plasma-Deposited Diamond", Science 297, 1670 (2002).

[4]A. R. Krauss, O. Auciello, M. Q. Ding, D. M. Gruen, Y. Huang, V. V Zhimov, E. I. Givarzigov, A. Breskin, R. Chechen, E. Shefer, V. Konov, S. Pimenov, A. Karabutov, A. Rakhimov, and N. Suetin, "Electron Field Emission for Ultrananocrystalline Diamond Films", J. Appl. Phys. 89, 2958 (2001).

[5]R. Locher, C. Wild, N. Herres, D. Behr, and P. Koidl, "Nitrogen Stabilized <100> Texture in Chemical Vapor Deposited Diamond Films", Appl. Phys. Lett., 65 [1] 3106 (1994).

[6]S. Koizumi, T. Teraji, and H. Kanda, "Phosphorus-Doped Chemical Vapor Deposition of Diamond", Diamond and Related Materials, 9, 935 (2000).

[7]I. Sakaguchi, M. N-Gamo, Y. Kikuchi, E. Yasu, and H. Haneda, "Sulfur: A Donor Dopant for n-type Diamond Semiconductors", Phys. Rev. B 60, R2139 (1999).

[8]S. Koizumi, "Growth and Characterization of Phosphorous Doped n-type Diamond Thin Films", Phys. Stat. Sol. (a) 172, 71 (1999)

[9]S. Battacharyya, O. Auciello, J. Birrell, J. A. Carlisle, L. A. Curtiss, A. N. Goyote, D. M. Gruen, A. R. Krauss, J. Schlueter, A. Sumant, and P. Zapol, "Synthesis and Characterization of Highly-Conducting Nitrogen-Doped Ultrananocrystalline Diamond Films", Appl. Phys. Lett.,79 [10] 1441 (2001).

[10]P. Zapol, M. Sternberg, L. A. Curtiss, T. Frauenhein, and D. M. Gruen, "Tight-binding molecular-dynamics simulation of impurities in ultrananocrystalline diamond grain boundaries", Phys. Rev. B 65, 045403 (2002).

[11]R. F. Mamin and T. Inushima, "Conductivity in Boron-Doped Diamond", Phys. Rev. B 63, 033201 (2001).

[12]H. Chatei, J. Bougdira, M. Remy, P. Alnot, C. Bruch, and J. K. Kruger, "Effect of Nitrogen Concentration on Plasma Reactivity and Diamond Growth in a H_2-CH_4-N_2 Microwave Discharge", Diamond and Related Materials, 6, 107-119 (1997).

[13]S. Bohr, R. Habner, and B. Lux, "Influence of Nitrogen on Hot-Filament Chemical Vapor Deposition of Diamond", Appl. Phys. Lett., 68, 1075 (1996).

[14]S. Jin and T. D. Moustakas, "Effect of Nitrogen on the Growth of Diamond Films", Appl. Phys. Lett., 65, 403 (1994).

[15]W. V. Smith, P. P. Sorokin, I. L. Gelle, and G. J. Lashe, "Electron-Spin Resonance of Nitrogen Donors in Diamond", Phys. Rev. 115, 1546 (1959).

[16]T. Vandevelde, M. Nesladek, C. Quaeyhaegens, and L. Stals, "Optical Emission Spectroscopy of the Plasma During CVD Diamond Growth with Nitrogen Addition", *Thin Solid Films*, 290-291, 143-147 (1996).

[17]A. Afzal, C. A. Rego, W. Ahmed, R. I. Cherry, "HFCVD Diamond Grown with Added Nitrogen: Film Characterization and Gas-Phase Composition Studies", Diamond and Related Materials7 (1998) 1033-1038.

[18]N. Wada, P. J. Gaczi, and S. A. Solin, "Diamond-like 3-fold Coordinated Amorphous Carbon", J. Non-Cryst. Solids 35, 543 (1980).

[19]F. L. Coffman, R. Cao, P. A. Pianetta, S. Kappor, M. Kelly, and L. J. Terminello, "Near-Edge X-Ray Absorption of Carbon Materials for Determining Bond Hybridization in Mixed sp^2/ sp^3 Bonded Materials", Appl. Phys. Lett. 69, 568 (1996).

[20]V. Shanov, W. Tabakoff, and R. N. Singh, "CVD Diamond Coatings for Erosion Protection at Elevated Temperatures" Journal of Materials Engineering and Performance, 11, 2 (2002).

[21]V. Shanov, R. N. Singh, and W. Tabakoff, "CVD Coatings for Erosion Protection at Elevated Temperatures" invited presentation at the International Conference on Metallurgical Coatings and Thin Films, April 10-14, 2000.

[22]Astex User Manual, HPM/M Magnetized HPMS Plasma Source, Version 1.6, Woburn, MA, 1992, p.5.

[23]R. Ramamurti, R. S. Kukreja, L. Guo. V. Shanov, R. N. Singh, "Microwave Plasma Chemical Vapor Deposition (CVD) of Carbon Based Films in the System C-N", Presented at the 28th International Conference and Exposition on Advanced Ceramics and Composites, Focused Session C: Nanomaterials and Biomimetics, January 25th – 30th, 2004, Cocoa Beach, Florida.

[24]Peak Fit 4.0 for Windows User's Manual, Peak Separation and Analysis Software, Copyright © 1997 by AISN Software (SPSS Inc., Chicago, IL, 1997).

[25]R. Ramamurti, V. Shanov, R. N. Singh, M. Samiee, and P. Kosel, edited by Y. Tzeng, K. Miyoshi, M. Yoshikawa, M. Murakawa, Y. Koga, K. Kobashi, and G. A. J. Amaratunga, Proceedings of the Sixth Applied Diamond Conference/ Second Frontier Carbon Technology (ADC/ FCT 2001), July, Auburn, Alabama, p.62 (2001).

[26]C. Kittel, "Introduction to Solid State Physics", 3rd Edition, John Wiley & Sons, Inc., New York, London, Sydney, 1966.

Porous Ceramics

MANUFACTURING OF HIGHLY-POROUS SIC-CERAMICS FROM SI-FILLED CELLULOSE FIBRE PAPERS

O. Rusina[1], R. Kirmeier[2], A. Molinero[2], C.R. Rambo[1] and H. Sieber[1]

[1]University of Erlangen-Nuremberg, Department of Materials Science, Glass and Ceramics, Erlangen, Germany,
[2]Paper Technology Specialist (PTS), Munich, Germany,

ABSTRACT
    Highly-porous, biomorphous SiC-ceramics were prepared by pyrolysis of Si-powder filled cellulose fibre materials in inert atmosphere at temperatures up to 1400°C. During annealing the Si-powder evaporated and reacted with the carbon of the pyrolysed cellulose fibres into β-SiC. After processing the highly-porous SiC-ceramics is characterized by a fibre-like morphology resulting from the initial cellulose fibre template.

INTRODUCTION
    Fabrication of porous, SiC-based ceramics and ceramic composites for applications as filter or catalyst support structures has attained particular interest in the last years [1-3]. Highly-porous SiC-ceramics were manufactured by the replication process [4,5], foaming of SiC containing slurries [6,7] or siliciding of carbon foams [8]. Recently, highly porous SiC ceramics were fabricated from biological derived preforms such as papers, cardboard or natural fibre textiles. Using different high-temperature infiltration-reaction processes, the bioorganic structures can be converted into ceramic composites within reasonable time [9-10]. The open-porous, biological structures are easily accessible for liquid or gaseous infiltrants during conversion. On the other hand, bioorganic materials are less expensive and exhibit excellent shaping and forming abilities. Si coated cellulose fibres [11,12] and Si_3N_4 coated cotton fibers [13] were converted into SiC-fibers by annealing in Ar-atmosphere at 1200-1600°C. Ohzawa et al. [14,15] infiltrated paper and cotton-cloth preforms with SiC by pressure-pulsed chemical vapor infiltration for high-temperature filter applications. Corrugated cardboard preforms [16,17] were used for infiltration with different preceramic polymer-filler suspensions and converted into ceramic composites.
    The aim of the present paper is to introduce the manufacturing of highly-porous SiC-ceramics by conversion of Si-powder filled cellulose fibre paper sheets. The Si-powder filled papers were produced by conventional paper processing technologies. The one step conversion processing includes the pyrolysis of the cellulose fibre paper into fibrous biocarbon templates and the subsequent high-temperature vapor-solid reaction by evaporation of the Si-powders in inert Ar-atmosphere into β-SiC.

EXPERIMENTAL PROCEDURE
    For preparation of the Si-powder filled paper sheets an unbleached kraft pulp was chosen as fiber raw material. This long-fibre cellulose provides a high paper strength, which is a requirement for paper grades with high filler contents. Due to the interference of the fillers on the fiber-fiber hydrogen-bonding strength losses may be caused. Cellulose fibers show a slightly anionic charge in water dispersion. Two different Si-powders (*Alfa AESAR, Germany*) with an average grain size of 3.6 μm and 6.2 μm were used as fillers.

An anionic starch was used both as retention aid and as an additive to improve the paper strength. Cationic additives were employed to adjust the charge balance and retain the metal particles onto the fibers. The dosage of each chemical additive was adapted for the different Si-filler contents. A nearly complete retention of Si-powder particles onto the fibers was achieved together with a good sheet formation and dewatering. The paper sheets showed an unexpected high strength. A water soluble latex was used together with cationic polymers to pre-flocculate the Si-powder particles in order to produce paper sheets with high filler contents and low decrease in the paper strength, formation and dewatering [18]. Si-filled paper sheets with an Si-content of 35, 45 and 70 wt% were prepared.

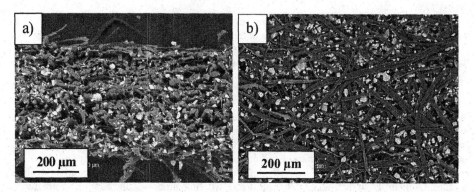

Figure 1: Mictrostructure of the initial Si-filled (35 wt.%) cellulose fiber paper, a) cross section, b) top-view.

Figure 1 shows SEM micrographs of a Si-powder filled cellulose fiber paper which contained 35 wt.% of Si-powder. The initial Si-filled paper shows a thickness of about 300 μm, Fig. 1a. The Si-powder particles between the cellulose fibers were distributed homogeneously over the cross section of the specimen paper sheet.

The Si-filled paper sheets were pyrolysed and the biocarbon from the cellulose fibers reacted with the Si-vapor into β-SiC at temperatures above 800 °C in Ar-atmosphere. Up to 500°C a slow heating rate of 5 K/min was applied to decompose the fiber polymers (cellulose, hemicellulose, lignin) into carbon [9-12, 19] during pyrolysis, followed by a faster heating rate of 10 K/min up to the peak temperature at 1400°C, Fig. 2.

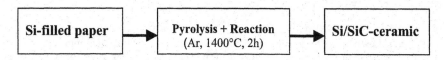

Figure 2: The processing scheme for preparation of SiC-ceramic from Si-filled paper.

The pyrolysis of the cellulose fibers and the Si powder / carbon reaction into β-SiC were studied by thermal balance analysis (TGA/DTA 409, *Netzsch Gerätebau, Selb, Germany*) and by X-ray diffraction using monochromatic CuK$_\alpha$ radiation (XRD, Diffrac 500, *Siemens, Mannheim, Germany*). Pore size and pore size distribution were measured using He-pycnometry and Hg-pressure analysis. The microstructure of initial and pyrolysed materials was characterized by scanning electron microscopy (SEM – XL30, *Fa. Philips, NL*).

RESULTS AND DISCUSSION

Fig. 3 shows the thermogravimetric analysis of the Si-filled cellulose paper in inert atmosphere. The TGA-scans show the decomposition of the Si-filled papers in two steps. In the first step the major decomposition reactions of cellulose and hemicellulose occurs, resulting in a rapid weight loss between 280 and 350°C [19]. At higher temperatures a smaller weight loss of about 10 wt.% is observed which may be attributed to the lignin decomposition. The total weight loss is almost terminated at 500-600°C indicating the presence of carbon only.

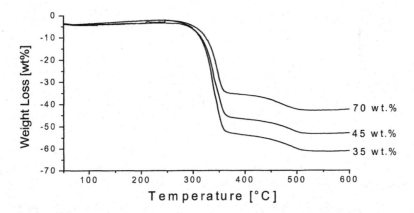

Figure 3: Weight loss of Si-powder filled paper with different Si-powder contents in Ar-atmosphere.

The reactive phase formation during annealing Si-powder filled paper sheets was analyzed by X-ray diffractomertry, Figure 4. The XRD-spectra show no phase formation at temperatures up to 1100°C. The formation of β-SiC as the ceramic product phase was observed at a temperature of 1400°C.

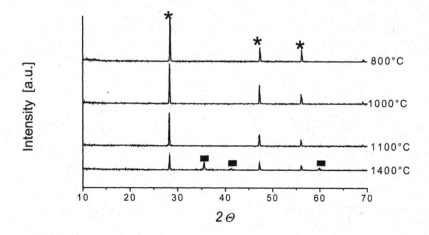

Figure 4: XRD spectra of the Si-filled (35 wt.%) paper after pyrolysis at different temperatures in Ar (*- Si, ∎– β-SiC).

The microstructure of Si-powder filled paper sheets after pyrolysis at 1100°C and after reaction of the carbonized cellulose fibers with the Si-vapor into β-SiC at 1400°C is shown in Figure 5. After pyrolysis, the initial cellulose fibers were transformed into carbon fibers with an average diameter of about 10 μm, Fig. 5b. The individual Si-powder particles are still located between the carbonized cellulose fibres. With increasing temperature above 1100°C the surface of the Si-powder particles shows a rounded morphology. Raising of the pyrolysis temperature to 1400°C resulted in a complete reaction of the carbonized cellulose fibers with Si-vapor into β-SiC. The morphology of the initial cellulose fibres was maintained forming a fibrous SiC-ceramic. The struts have a thickness ranging from 5 to 15 μm and pore diameters up to 50 μm.

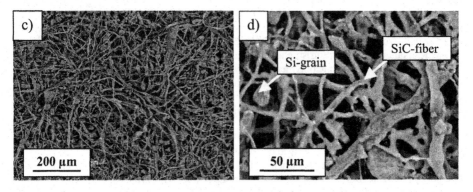

Figure 5: SEM micrograph of Si-filled (35%) paper: a)-b) after pyrolysis at 1100°C, c) and d) after pyrolysis and SiC-reaction at 1400°C in Ar.

CONCLUSIONS

Highly porous SiC-ceramics were produced from Si-filled cellulose paper sheets by pyrolysis in inert atmosphere. During the conversion of Si-powder filled paper into the ceramic a linear shrinkage of about 10 % was observed. After processing the highly-porous SiC-ceramics is characterized by a fiber-like morphology resulting from the initial cellulose fiber template. The conversion of Si-filled cellulose fiber papers into fibrillar SiC offer a new and low cost processing route for manufacturing of highly porous light weight SiC-ceramics.

ACKNOWLEDGEMENT

The financial support from the AiF Z0102, Volkswagen Foundation under contract # I/73 043 and helpful discussion with Prof. Peter Greil are gratefully acknowledged.

REFERENCES

[1]   J. Oi-Uchisawa, A. Obuchi, S. Wang, T. Nanba and A. Ohi, "Catalytic performance of Pt/MO$_x$ loaded over SiC-DPF for soot oxidation", *Applied Catalysis B: Environmental* **43** [2] 117-129(2003).

[2]   D. Fino, P. Fino, G. Saracco and V. Specchia, "Innovative means for the catalytic regeneration of particulate traps for diesel exhaust cleaning", *Chemical Engineering Science* **58** [3-6] 951-958(2003).

[3]   P. Pastila, V. Helanti, A.Nikkilä and T. Mäntylä, "Environmental effects on microstructure and strength of SiC-based hot gas filters", *Journal of the European Ceramic Society* **21** [9] 1261-1268 (2001).

[4]   X. Zhu, D. Jiang, Sh. Tan, "Preparation of silicon carbide reticulated porous ceramics", *Materials Science and Engineering* A**323**, 232-238 (2002).

[5]   T.J. Fitzgerald, V.J. Michaud, A. Mortensen, "Processing of microcellular SiC foams Part

II Ceramic foam production", *Journal of Materials Science*, **30**, 1037-1045 (1995).

[6] Y. Aoki, B. McEnaney, "SiC foams produced by siliciding carbon foams", *British Ceramic Transactions*, **94** [4] 133-137 (1995).

[7] M.R. Nangrejo, M.J. Edirisinghe, "Porosity and strength of silicon carbide foams prepared using preceramic polymers", *Journal of Porous Materials*, **9**, 131-140 (2002).

[8] P. Colombo, J.R. Hellmann, "Ceramic foams from preceramic polymers", *Mat. Res. Inn.* **6** (2002) 260.

[9] P. Greil, "Biomorphic ceramics from lignocellulosics", *J. Eur. Ceram. Soc.* **21** 105(2001).

[10] H. Sieber, A. Kaindl, D. Schwarze, J.-P. Werner and P. Greil, "Light-weight cellular ceramics from biologically-derived performs", *cfi/Ber. DKG* **77** 21(2000).

[11] H. Sieber, A. Kaindl , H. Friedrich and P. Greil, *"Crystallization of SiC on biological carbon precursors"*, Ceramic Transactions 110 *(Bioceramics: Materials and Applications III)*, Ed.: L. George, R. P. Rusin, G. S. Fischman and V. Janas, The American Ceramic Society, pp. 81-92 (2000).

[12] T. Fey, H. Sieber and P. Greil, "Manufacturing of highly porous SiC-ceramics from cellulose fibres", Proceedings of 5th International Conference on High Temperature Ceramic Matrix Composites (HTCMC-5), Seattle, WA/USA 2004, in print.

[13] R.V. Krishnaro, Y.R. Mahajan, "Preparation of silicon carbide fibres from cotton fibre and silicon nitride", *J. Mat. Sci. Lett.* **15**, 232 (1996).

[14] Y. Ohzawa, A. Sadanaka and K. Sugiyama, "Preparation of gas-permeable SiC shape by pressure-pulsed chemical vapour infiltration into carbonized cotton-cloth performs", *J. Mat. Sci.* **33** 1211(1998).

[15] Y. Ohzawa, H. Hshino, M. Fujikawa, K. Nakane and K. Sugiyama, "Preparation of high-temperature filter by pressure-pulsed chemical vapour infiltration of SiC into carbonized paper-fibre performs", *J. Mat. Sci.* **33** 5259 (1998).

[16] P. Greil, H. Sieber, D. Schwarze and H. Friedrich, *"Manufacturing of light weight Ceramics from cellulose structures"*, Ceramic Transactions 112 *(Ceramic Processing Science VI)*, Ed.: S. Hirano, G.L. Messing and N. Claussen, The American Ceramic Society, 527 - 532 (2001).

[17] H. Sieber, D. Schwarze, A. Kaindl , H. Friedrich and P. Greil, *"Ceramic lightweight structures from paper derived composites"*, Ceramic Transactions 108 *(Innovative Processing and Synthesis of Ceramics, Glasses and Composites III)*, Ed.: J.P. Singh, N.P. Bansal and K. Niihara, The American Ceramic Society, pp. 571-580 (2000).

[18] Bobalek E.G., Mendoza A.P., "Compositing latex polymers with pulp fibre", *Pulp and Paper Canada*, 72 [9] 95 (1971).

[19] D. Klemm, B. Philipp, T. Heinze, U. Heinze, W. Wagenknecht, "Comprehensive Cellulose Chemistry", *Vol. I, Wiley-VCH, Weinheim* 107(1998).

Kinetics and Mechanism

HIGH TEMPERATURE FORMING OF CERAMIC SHAPES WITHOUT APPLYING EXTERNAL PRESSURE

Stephen J. Lombardo, Rajesh V. Shende, and Chang Soo Kim
Department of Chemical Engineering
University of Missouri
Columbia, MO 65211

Robert A. Winholtz
Department of Mechanical and and Aerospace Engineering
University of Missouri
Columbia, MO 65211

ABSTRACT

Thin ceramic bodies have been shaped into more complicated shapes at elevated temperature without the application of an external pressure. To achieve deformation, thin beams of alumina have been coated with magnesia; at high temperature, the strain mismatch which arises in the sample leads to deformation. X-ray diffraction has shown that a new phase, spinel, forms, and it is presumably the volume change associated with the phase change that drives the deformation. The magnitude of the strain mismatch driving the deformation is quite large, and models for describing such large curvature in terms of the sample and coating thicknesses, Young's moduli, and strain mismatch are presented here.

INTRODUCTION

In earlier work, we have demonstrated a method to shape thin ceramic components by first coating substrates with a powder and then heating the coated substrates to elevated temperature [1-3]. This process was first demonstrated [1,2] by coating dense alumina beams with a layer of chromia powder. Upon heating, the substrates, which were initially flat, deformed into the shape of an arc. The substitution of larger chromium cations for the smaller aluminum cations in the corundum structure was suggested as the driving force for deformation.

More recently, we have extended the methodology by coating green alumina substrates with magnesium oxide powder [3]. This modified process has several advantages. First, the green alumina substrates undergo significantly more deformation than dense substrates. Secondly, the use of chromia is disadvantageous because of its volatility. The use of magnesia coatings on alumina also leads to a new driving force for deformation, namely, the formation of a second phase, which in this case is spinel.

In this work, we first present some results for the large curvature achieved in samples shaped by this forming method. During development of this forming method, a number of questions arose as to the mechanism of the deformation in such biceramic or functionally gradient materials. The deformation behavior of such components has received much attention in the literature [4-12], and this work applies analytical models of linear elastic and viscoplastic mechanics to account for the observed level of deformation.

EXPERIMENTAL

The process used to deform the substrates is shown schematically in Fig. 1. A slurry containing 84 wt% A16-SG powder (Alcoa, Bauxite, AR) in 16 wt% water was prepared with a dispersant (Duramax D 3005, Rohm and Hass, PA). The slurry was milled for 24 h using cylindrical alumina beads and then filtered. Binder (Duramax B1035, Rohm and Hass, PA) at 10 wt% was next added under low shear. Green tapes over a range of thickness were then prepared by tape casting. After drying, the tapes were cut into specimens.

To coat the green substrates, a water-based paint of magnesia powder was applied followed by drying in air at 100°C for 30 min. The magnesia uptake was determined by the difference in sample weight before and after coating, and includes the binder. The coated substrates were placed on zirconia setters with the coated-side on top and were heated in a combined sintering/forming cycle in a box furnace in air with a heating rate of 10°C/min to the soak temperature, followed by a 4 h hold at soak temperatures between 1500 and 1700°C. After firing, the dimensions and geometry of the samples (see Fig. 1) were used to determined the curvature, κ , which is the reciprocal of the radius of curvature, R. When no deformation occurs, both segment height A and κ are zero.

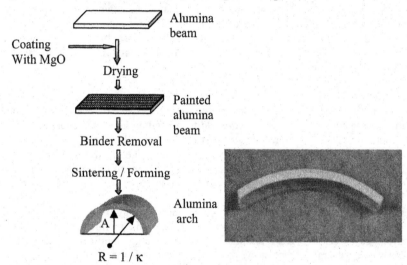

Fig. 1 Left) Process used to deform a flat green beam into an arc of curvature, κ. The segment height, A, is also a measure of the amount of deformation. Right) Curved substrate with κ= 0.05 mm^{-1} and R=20 mm.

RESULTS AND DISCUSSION

Figure 2 shows the degree of curvature of samples heated at temperatures between 1500-1700°C. In general, the curvature exhibited by samples fabricated with this method is fairly uniform and quite large. We are thus interested in knowing the magnitude of the strain mismatch that drives the deformation and the mechanics that can account for such large deformations. A number of models have appeared in the literature for describing the deformation in such systems [4,6,8,11,12], and here we show some relationships between the different mechanics models, namely the Stoney model, a viscoplastic model, and the bilayer, e.g., bimetallic, strip model.

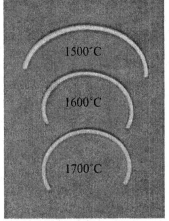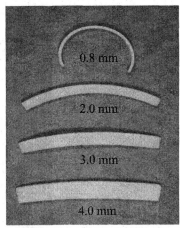

Fig. 2 Left) Curved substrates of 0.8 mm thickness formed from initially flat alumina substrates coated with magnesia and then held at different soak temperatures for 4 h. Right) Curved substrates of different thickness formed from initially flat alumina substrates coated with magnesia and then held at 1600°C for 4 h.

Figure 3 shows a schematic for pure bending in a beam, initially of length L_o, driven by a strain, ε_1, in the top layer. This strain thus increases the length of the top layer of the beam to L_1, which is given by

$$L_1 = L_o + \varepsilon_1 L_o \qquad\qquad 1$$

or, equivalently,

$$\varepsilon_1 = \frac{L_1 - L_o}{L_o} \qquad\qquad 2$$

For any position within the deformed beam, the strain can be determined from the compatibility equations which leads to

$$\varepsilon = \varepsilon_o + \kappa z \qquad\qquad 3$$

where ε_o is an arbitrary strain in the body at location $z=0$, which in this case corresponds to the interface. In the deformed beam, a point of zero strain exists at the skeletal point, z_{sk}.

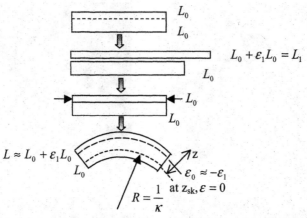

Fig. 3 Schematic of strain mismatch in a beam leading to deformation. The top layer of the beam is first allowed to expand to L_1, and then is compressed back to the original length, L_0. Finally, the composite beam is allowed to bend.

For bodies which undergo large curvature, plastic deformation may govern the deformation of the body. The relation between the original strained length of the top surface, L_1, and the length in the deformed shape, L, can thus be obtained from Eq. 3 as

$$\frac{L}{L_1} = \frac{1 + \kappa z_{sk}}{1 + \varepsilon_1} \tag{4}$$

For small curvatures and strains, $L \approx L_1$ and thus $\varepsilon_0 \approx -\varepsilon_1$. With this relationship between the two strains, the curvature can be expressed as

$$\kappa = \frac{-\varepsilon_o}{z_{sk}} \approx \frac{\varepsilon_1}{z_{sk}} \tag{5}$$

Stoney Problem

The Stoney problem treats a thin uniformly strained layer (ε_1 = constant) on a substrate [4]. A schematic of the system is shown below and tension is taken as positive and compression as negative. The location $z=0$ coincides with the interface between the two layers, which is the position of the x axis, and the location $z=z_{sk}$ is the point of zero strain in the substrate.

The stresses within the beam are given by

$$\sigma_1 = E_1 \varepsilon_1 \tag{6a}$$
$$\sigma_2 = E_2 \varepsilon_2 = E_2 \kappa (z - z_{sk}) \tag{6b}$$

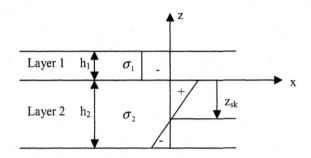

Fig. 4 Schematic of the stress distribution and the skeletal point, z_{sk}, for a bilayer beam undergoing deformation

Equilibrium of forces, $\sum F = 0$, leads to

$$\int_{-h_2}^{0} \sigma_2 dz + \int_{0}^{h_1} \sigma_1 dz = 0 \qquad\qquad 7$$

In light of Eq. 6b, this can be expressed as

$$\int_{-h_2}^{0} E_2 \kappa (z - z_{sk}) dz + \int_{0}^{h_1} \sigma_1 dz = 0 \qquad\qquad 8$$

which leads to

$$-\frac{E_2 \kappa h_2^2}{2} - E_2 \kappa z_{sk} h_2 + \sigma_1 h_1 = 0 \qquad\qquad 9$$

In a similar fashion, equilibrium of moments, $\sum M = 0$, leads to

$$\int_{-h_2}^{0} \sigma_2 z dz + \int_{0}^{h_1} \sigma_1 z dz = 0 \qquad\qquad 10$$

In the classic Stoney problem [4], the bending moment in the film is neglected because the film is treated as being infinitesimally small, and thus neglect of the term in σ_1 in Eq. 10 leads to

$$E_2 \kappa \left[-\frac{z_{sk} h_2^2}{2} - \frac{h_2^3}{3} \right] = 0 \qquad\qquad 11$$

where the stresses in Eq. 10 have been replaced by Eq. 6b. The location of the skeletal point is thus

$$z_{sk} = -\frac{2}{3}h_2 \qquad\qquad 12$$

and is independent of the elastic moduli of the two layers and the thickness, h_1. When Eq. 12 is combined with Eq. 9 and rearranged, we obtain

$$\kappa = -\frac{6\sigma_1 h_1}{E_2 h_2^2} \qquad\qquad 13$$

This is the result obtained by Stoney [4], which expresses the curvature in terms of the properties of the two layers. If the stress in Eq. 13 is replaced by Eq. 6a, then we obtain

$$\kappa = -\frac{6E_1 h_1}{E_2 h_2^2}\varepsilon_1 \qquad\qquad 14$$

The above approach can be extended for thicker coating layers if the bending moment in the beam is taken into account explicitly in the summation of the moments [6,8,12]. Starting from Eq. 10 combined with Eqs. 6a and 6b, we obtain for the summation of the bending moments

$$\int_{-h_2}^{0} E_2 \kappa (z - z_{sk}) z\,dz + \int_{0}^{h_1} E_1 \varepsilon_1 z\,dz = 0 \qquad\qquad 15$$

which leads to

$$E_2\left(\frac{\kappa h_2^3}{3} + \frac{\kappa z_{sk} h_2^2}{2}\right) + \frac{E_1 \varepsilon_1 h_1^2}{2} = 0 \qquad\qquad 16$$

In light of Eq. 9, this can be expressed as

$$E_2\left(\frac{h_2^3}{3}\kappa + \frac{z_{sk} h_2^2}{2}\kappa\right) + \left(\frac{E_2 h_2^2}{2h_1}\kappa + \frac{E_2 h_2 z_{sk}}{h_1}\kappa\right)\frac{h_1^2}{2} = 0 \qquad\qquad 17$$

Equation 17 can be rearranged to find the location of the skeletal point as

$$z_{sk} = -h_2 \frac{\dfrac{2}{3} + \dfrac{h_1}{2h_2}}{1 + \dfrac{h_1}{h_2}} \qquad\qquad 18$$

which is independent of the elastic moduli of the two layers. When Eqs. 18 and 6a are substituted into Eq. 9, we obtain

$$\kappa = -\frac{6E_1 h_1}{E_2 h_2^2}\left(\frac{h_1}{h_2}+1\right)\varepsilon_1 \qquad\qquad 19$$

A comparison of Eqs. 14 and 19 indicates that the error in κ between the two equations is proportional to $(1+h_1/h_2)$. Thus if h_1 is less than 10-20% of h_2, then error of the same amount is introduced by using the original result, Eq. 14.

For cases where the body undergoes large deformation beyond the elastic limit of the materials, Eq. 5 can be used with Eq. 18 for z_{sk} to describe the curvature as

$$\kappa = \frac{\varepsilon_1}{z_{sk}} = -\frac{1+\dfrac{h_1}{h_2}}{\left(\dfrac{2}{3}+\dfrac{h_1}{2h_2}\right)}\frac{\varepsilon_1}{h_2} \qquad\qquad 20$$

This is the modified result for curvature in terms of the strain in the film. If $h_1 \ll h_2$, then the curvature at the skeletal point in the thin film limit is

$$\kappa = -\frac{3\varepsilon_1}{2h_2} \qquad\qquad 21$$

which is independent of the thickness of the uniformly strained layer, h_1. Equation 21 can also be obtained directly from Eqs. 5 and 12. We note that Eqs. 14 and 21 have different mathematical forms for the curvature, which were derived based on different views for the nature of the deformation mechanics. We further note that both Eqs. 14 and 21 can be obtained from the model for a bilayer strip, and this is shown below.

Bimetallic Strip
A schematic of the geometry of the bimetallic strip is given in Fig. 5 and tension is taken as positive and compression as negative. In each layer, the elastic stresses are:

$$\sigma_1 = E_1(\varepsilon - \alpha_1 \Delta T) \qquad\qquad 22a$$
$$\sigma_2 = E_2(\varepsilon - \alpha_2 \Delta T) \qquad\qquad 22b$$

where α_i is the CTE value for material i.

Equilibrium of forces and moments leads to the curvature at the interface as [6]

$$\kappa = \frac{6\left(E_1 E_2 h_1 h_2 \Delta T (\alpha_1 - \alpha_2)(h_1 + h_2)\right)}{E_1^2 h_1^4 + E_1 E_2 h_1 h_2 (4h_1^2 + 6h_1 h_2 + 4h_2^2) + E_2^2 h_2^4} \qquad\qquad 23$$

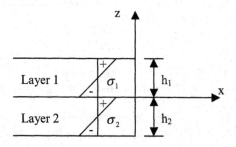

Fig. 5. Schematic of the stress distribution and coordinate system for a bilayer beam undergoing deformation.

If we define $D = (\alpha_1 - \alpha_2)\Delta T$, then Eq. 23 can be rearranged to

$$\kappa = \frac{6\left(\dfrac{E_1}{E_2}\right)\left(\dfrac{1}{h_2}\right)\left(\dfrac{h_1}{h_2}\right)\left(\dfrac{h_1}{h_2}+1\right)}{\left(\dfrac{E_1}{E_2}\right)^2\left(\dfrac{h_1}{h_2}\right)^4 + \left(\dfrac{E_1}{E_2}\right)\left(\dfrac{h_1}{h_2}\right)\left[4\left(\dfrac{h_1}{h_2}\right)^2 + 6\left(\dfrac{h_1}{h_2}\right)+4\right]+1}D \qquad\qquad 24$$

In the thin film limit of $h_1 < h_2$, neglect of higher order terms in h_1/h_2 leads to

$$\kappa = \frac{6\dfrac{E_1}{E_2}\dfrac{h_1}{h_2^2}}{4\dfrac{E_1}{E_2}\dfrac{h_1}{h_2}+1}D \qquad\qquad 25$$

For specified relationships between E_1 and E_2, Eq. 25 simplifies to the forms of the two equations derived earlier, namely Eqs. 14 and 21. If $E_1/E_2 << h_2/4h_1$, then we obtain Eq. 14. Conversely, if $E_1/E_2 >> h_2/4h_1$, then we obtain Eq. 21.

To compare a number of different cases for the curvature predicted by the different models, we define dimensionless modulus, thickness ratio, and curvature as

$$\alpha = \frac{E_1}{E_2} \qquad\qquad 26$$

$$\beta = \frac{h_1}{h_2} \qquad\qquad 27$$

$$\gamma = \frac{h_2\kappa}{D} \qquad\qquad 28$$

where we are equating $D = -\varepsilon_1$ as the strain mismatch in the sample. Equation 24 therefore becomes

$$\gamma = \frac{6\alpha\beta(\beta+1)}{\alpha^2\beta^4 + \alpha\beta\left(4\beta^2 + 6\beta + 4\right) + 1} \qquad\qquad 29$$

Equation 21 is obtained for $\alpha\beta > 1/4$ and Eq. 14 results for $\alpha\beta < 1/4$. Figure 6 shows a comparison of the dimensionless curvature, γ, predicted by the different models as a function of the dimensionless thickness ratio, β. For large deformations (Eq. 21, $\gamma = 3/2$ in dimensionless form), a constant curvature is predicted which is independent of the thickness ratio and modulus ratio of the two layers. The Stoney model (Eq. 14, $\gamma = 6\alpha\beta$ in dimensionless form) depends on both the moduli and thickness ratios and is a family of curves parametric in α, the modulus ratio. The full bilayer model (Eq. 24 and Eq. 29 in dimensionless form) also depends on both the modulus and thickness ratios, but now in a more complicated way than for the Stoney model. We also can clearly see the range of accuracy of the Stoney model, which is tangential at low β to the corresponding curve from Eq. 29 for the same value of the modulus ratio. For very large values of the modulus ratio, e.g., $\alpha > 10$, a plateau region of curvature appears which is independent of the thickness of h_1. The maximum value of the curvature of the plateau region predicted from Eq. 29 is given by the relation $\gamma = 3/2$.

Figure 6 also shows data points for the samples fabricated in this work at different temperatures between 1500-1700°C. The value of h_1 was determined by EDS profiling of the degree of Mg penetration into the sample, and h_2 was determined from h_1 and the measured thickness of the sample. The moduli ratio, α, and strain mismatch, D, were then determined by nonlinear regression analysis. For the samples fabricated in this work, D ranges from 0.045-0.06 and α ranges from 10-20. For the samples fabricated at thickness ratios above $\beta = 0.05$, the observed curvature is within 20% of the upper limit curvature of $\gamma = 3/2$. Thus, based on the trends in Fig. 6, the magnitude of the curvature observed in this work can be qualitatively explained by either Eq. 21 or 29.

If Eq. 29 is used to describe the magnitude of the curvature, then a rather large value of the modulus ratio $\alpha = 10-20$ must be used. The values of the Young's moduli for dense alumina and spinel are 380 GPa and 250 GPa respectively, and thus in no way approach a value as high as 10-20. During the period under which deformation occurs, however, the relative density of the two phases may be substantially different as the spinel phase forms and sinters and the alumina phase sinters. Because the modulus of a material is a strong function of density, especially at intermediate densities, this may partially account for the large modulus ratio, especially if the spinel phase forms and sinters more rapidly than alumina sinters. Alternatively, if we think of the modulus ratio E_1/E_2 as $\sigma_1/(\varepsilon_1 E_2)$, then although the strain and modulus remain finite, the stress may become quite large as atom substitution occurs in the formation of the new phase.

Alternatively, if Eq. 21 is used to account for the deformation, the curvature arises because of a large mismatch strain between the substrate and the new phase formed in the thin top layer of the substrate. For the samples fabricated here, a strain mismatch of approximately 0.05 is required to realize such large curvatures. For ceramics undergoing elastic strain, such large deformations without fracture are not possible. At elevated temperature, however, where viscoplastic deformation can occur, such large strains are possible. The data in Figure 6 may thus suggest that both linear elastic and viscoplastic mechanics are operative. For the samples fabricated at the temperature of 1500°C, the curvature exhibits the largest dependence on the thickness of the top layer, as is predicted in Eq. 29; such behavior may denote a larger contribution of linear elastic mechanics. For the samples fabricated at the highest temperature of 1700°C, the curvature is less dependent on thickness of the top layer, as is predicted in Eq. 21, which may be indicative of a larger contribution of viscoplastic mechanics.

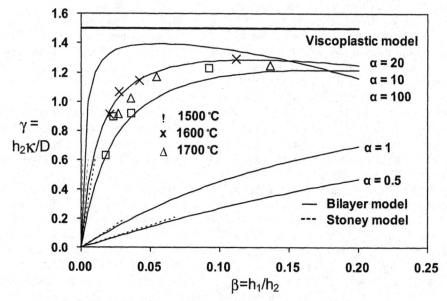

Fig. 6 Comparison of the dimensionless curvature versus thickness ratio and modulus ratio predicted by the different models corresponding to equations 14, 21, and 29. The curvature for samples fabricated at different temperatures is also indicated.

CONCLUSIONS

A method has been developed to introduce large amounts of strain mismatch into substrates without applying an external pressure. The method is based upon coating a green substrate of alumina with magnesia powder and then heating the composite beam to high temperature. The volume change associated with the formation of the spinel phase is the driving force for deformation.

The degree of deformation exhibited by substrates formed by this method can be large, with associated strain mismatches of up to 5%. The degree of deformation can be accounted for in terms of the thickness of the two layers, the strain mismatch, and the ratio of the Young's moduli.

ACKNOWLEDGMENT

Partial support for this work was from the National Science Foundation under Grant No. 0203136 and by the University of Missouri Research Reactor (MURR).

REFERENCES

1. S. J. Lombardo, D. Bianchi, B. Bishop, A. Giannakopoulos, R. Goldsmith, R. Higgins, R. Pober, S. Suresh, "Forming of Ceramics during Firing without the Application of External Pressure," J. Am. Ceram. Soc. **82** 1401-1408 (1999).
2. S. J. Lombardo, A. Giannakopoulos, S. Suresh, D. Bianchi, B. Bishop, R. Goldsmith, R. Higgins, and R. Pober, "Shaping of Ceramics at Elevated Temperature without External Applied Pressure," Eds. J. P. Singh, N. P. Bansal, K Niihara, in *Ceramic Transactions*, Vol. 108 Innovative Processing and Synthesis of Ceramics, Glasses, Composites III (American Ceramic Society, Westerville, OH, 2000) 213-221.
3. S. J. Lombardo, R. V. Shende, C. S. Kim, and R. A. Winholtz, "Strain-Induced Forming of Ceramics Without the Application of External Pressure," in *Ceramic Transactions* Vol. 154, Innovative Processing and Synthesis of Ceramics, Glasses, and Composites, VII (American Ceramic Society, Westerville, OH, 2003) 115-121.
4. G. G. Stoney, "The Tension of Metallic Films Deposited by Electrolysis," Proc. R. Soc. London, A, **82** 172-175 (1909).
5. Y. Sugawara, K. Onitsuka, S. Yoshikawa, Q. Xu, R. E. Newnham, and K. Uchino, "Metal-Ceramic Composite Actuators," J. Am. Ceram. Soc. **75** 996-998 (1992).
6. S. Suresh, A. E. Biannakopoulos, and M. Olsson, "Elastoplastic analysis of Thermal Cycling; Layered Materials with Sharp Interfaces," J. Mech. Phys. Solids, **42** 979-1018 (1994).
7. Y.-L. Shen and S. Suresh, "Thermal Cycling and Stress Relaxation Response of Si-Al and Si-Al-SiO$_2$ Layered Thin Films," Acta Metall. Mater., **43** 3915-3926 (1995).
8. L. B. Freund, "Some Elementary Connections between Curvature and Mismatch Strain in Compositionally Graded Thin Films," J. Mech. Phys. Solids, **44** 723-736 (1996).
9. W.Y. Shih, W.-H. Shih, and I. H. Aksay, "Scaling Analysis for the Axial Displacement and Pressure of Flextensional Transducers," J. Am. Ceram. Soc. **80** 1073-1078 (1997).
10. G. Li, E. Furman, and G. H. Haertling, "Stress-Enhanced Displacements in PLZT Rainbow Actuators," J. Am. Ceram. Soc. **80** 1382-1388 (1997).
11. L. B. Freund, "Substrate Curvature due to Thin Film Mismatch Strain in the Nonlinear Deformation Range," J. Mechanics & Physics of Solids, **48** 1159-1174 (2000).
12. C. A. Klein, "How Accurate are Stoney's Equation and Recent Modifications," J. Appl. Physics, **88** 5487-5489 (2000).

THE INFLUENCE OF BINDER DEGRADATION KINETICS ON RAPID BINDER REMOVAL CYCLES

Stephen J. Lombardo and
Jeong Woo Yun
Department of Chemical Engineering
University of Missouri
Columbia, MO 65211, USA

Daniel S. Krueger
Honeywell
Federal Manufacturing
& Technologies, LLC
Kansas City, Missouri 64141, USA

ABSTRACT
In order to predict minimum time heating cycles for binder removal, an accurate representation of the kinetics of binder decomposition is needed. In this work, we show how the activation energy and preexponential factor for the kinetics of binder decomposition can be determined. We examine some issues related to the uniqueness of such kinetic constants and their overall accuracy in describing weight loss profiles determined by thermogravimetric analysis. Once the kinetics have been accurately determined, they can then be used to determine the minimum time heating cycles for binder removal from porous green ceramic bodies.

INTRODUCTION
The specification of heating cycles for the thermal removal of binder [1-12] from porous green ceramic bodies is often a trial-and-error procedure. We have recently developed an analytical model to predict the minimum cycle time for the removal of binder from green ceramic bodies when open porosity exists [13]. Under such circumstances, the exiting of the gas products of binder degradation is governed by flow in porous media. The minimum cycle time for the removal of binder then depends on a number of transport and kinetic quantities such as the dimensions of the green body, the permeability of the green body, the volume fraction of binder, and the rate of binder degradation.

In this study, we examine how the kinetics of binder degradation [14,15] can be determined from weight loss data as obtained by thermogravimetric analysis (TGA). For many real systems of binder and ceramic, the decomposition behavior is complex and may not be straightforward to interpret. Some of the issues related to analyzing TGA data and to the accuracy of different methods for determining the activation energy and preexponential factor for binder decomposition are discussed here.

THEORY
We first note that the time, t, required to remove binder from an initial volume fraction of binder, $\varepsilon_{b,o}$, to some other volume fraction of binder, ε_b, can be represented as [12,13]

$$t = \frac{GT_s}{P_t^2 - 1} \int_{\varepsilon_b}^{\varepsilon_{b,o}} \frac{d\varepsilon_b}{\kappa(\varepsilon_b)} \qquad\qquad 1$$

where the symbols are defined in the Nomenclature Section and the quantity G is given as

$$G = 0.8365 \frac{\mu \rho_b}{2\rho_o^2 RMT_o^2} \frac{L_x^2 L_y^2 L_z^2}{L_x^2 L_y^2 + L_x^2 L_z^2 + L_y^2 L_z^2} \qquad 2$$

where L_x, L_y, and L_z are the dimensions of the parallelepiped body The minimum cycle time, t^*, is then obtained for the complete removal of binder where the lower limit of integration in Eq. 1 is set at $\varepsilon_b = 0$ for $t = t^*$:

$$t^* = \frac{GT_s}{P_t^2 - 1} \int_0^{\varepsilon_{b,o}} \frac{d\varepsilon_b}{\kappa(\varepsilon_b)} \qquad 3$$

To complete the model description, we represent the permeability, κ, in Eqs. 1 and 3 by the Kozeny-Carmen equation as

$$\kappa = \frac{\varepsilon^3}{k(1-\varepsilon)^2 S^2} \qquad 4$$

where k is a constant and S is the surface area per unit volume. When Eq. 4 is combined with Eq. 3, the minimum cycle time can be represented in closed form as [13]

$$t^* = \frac{GT_s k S^2}{P_t^2 - 1} \times$$
$$\left\{ \ln \frac{1-\varepsilon_s}{1-\varepsilon_s - \varepsilon_{b,o}} - 2\left[\frac{1}{1-\varepsilon_s - \varepsilon_{b,o}} - \frac{1}{1-\varepsilon_s} \right] + \frac{1}{2}\left[\frac{1}{(1-\varepsilon_s - \varepsilon_{b,o})^2} - \frac{1}{(1-\varepsilon_s)^2} \right] \right\} \qquad 5$$

Equation 5 can thus be used to determine the minimum time for binder removal when the dimensional and transport parameters have been specified. It can be observed that the minimum time, t^*, in Eq. 5 does not explicitly depend on kinetic parameters such as the activation energy, E, and the preexponential factor, A. As will be seen shortly, however, T_s, the starting temperature for the minimum time heating cycle that appears in Eq. 5, does depend on the values of A and E and on the form of the kinetic expression.

Equation 5 describes the length of the minimum time heating cycle but does not indicate how the temperature should be varied as a function of time. To find this, we first represent the binder decomposition kinetics in an Arrhenius form as [14,15]

$$r = \frac{d\alpha}{dt} = A\exp(-\frac{E}{RT})f(\alpha) \qquad 6$$

where the function $f(\alpha)$ describes how the concentration or volume fraction of binder influences the degradation kinetics. Many forms of $f(\alpha)$ have been proposed, and a brief summary is given in Table I. The first three mechanisms in Table I describe zero-, first-, and second-order decomposition processes whereas the last two mechanisms describe degradation kinetics controlled by diffusion [14-18].

To obtain values for A and E in Eq. 6 from TGA weight loss data, Eq. 6 can be integrated between two times, t_1 and t_2, which correspond to two temperatures, T_1 and T_2, and to two degrees of conversion, α_1 and α_2. Because the exponential function in Eq. 6 cannot be integrated analytically, numerical methods or approximations to the integral must be used. An approximate method of high accuracy has been developed by Lee and Beck [19] to determine A and E from TGA weight loss data obtained with a linear heating rate, β:

$$\ln\left[\frac{F(\alpha)}{T^2}\right] = \ln\left[\frac{AR}{\beta(E+2RT)}\right] - \frac{E}{RT} \qquad\qquad 7$$

The function $F(\alpha)$ in Eq. 7 is the integrated form of $f(\alpha)$ given by

$$F(\alpha) = \int \frac{1}{f(\alpha)}\,d\alpha \qquad\qquad 8$$

Expressions for $F(\alpha)$ corresponding to the different kinetic models $f(\alpha)$ are also given in Table I. To obtain values of A and E, $F(\alpha)$ is computed from weight loss data at each temperature and then the left-hand-side of Eq. 7 is plotted versus $1/T$ [17,18]. For a full or partial range of conversion, a linear region will be evident with slope $-E/R$. This value of E can then be used with Eq. 7 to determine A.

Table I. Kinetic mechanisms of polymer decomposition, f, and their integrated forms, F, in terms of the fractional conversion, α, where $\alpha=1-\varepsilon_b/\varepsilon_{bo}$.

Kinetics	$f(\alpha)$	$F(\alpha)$
0 order	$(1-\alpha)^0$	α
1st order	$(1-\alpha)^1$	$-ln(1-\alpha)$
2nd order	$(1-\alpha)^2$	$\alpha/(1-\alpha)$
Diffusion A	$[-ln(1-\alpha)]^{-1}$	$(1-\alpha)\,[ln(1-\alpha)]+\alpha$
Diffusion B	$3(1-\alpha)^{1/3}/[2(1-\alpha)^{-1/3}-1]$	$[1-(1-\alpha)^{1/3}]^2$

When the kinetic quantities A, E, and $f(\varepsilon_b)$ have been determined by analyzing TGA weight loss data as indicated above, the manner in which the temperature $T(\varepsilon_b)$ should be varied for the binder removal cycle can then be expressed in terms of these quantities as [13]

$$T(\varepsilon_b) = \frac{-E}{R}\left[\ln\frac{\kappa}{f(\varepsilon_b)}\frac{(P_t^2-1)}{GT_sA}\right]^{-1} \qquad\qquad 9$$

Equation 9 is an algebraic relationship between T and ε_b. Because Eq. 1 provides a mathematical relationship between t and ε_b, and Eq. 9 provides a relationship between T and ε_b, the relationship between T and t, namely the heating cycle, can thus be determined. We note that because $T(\varepsilon_b)$ in Eq. 9 depends A, E, and $f(\varepsilon_b)$, these quantities will influence the heating cycles obtained from Eqs. 1 and 9. It is therefore the aim of this work to indicate how the assumption of a kinetic mechanism and the analysis of TGA data influences the values of A and E, which are then used to specify the minimum time heating cycle for binder removal.

RESULTS

In earlier work [16-18], we have shown how to take TGA data for real binder systems and obtain values of A and E that can adequately represent the kinetics of decomposition. In general, the degradation of binder is a complex process, and the assumption of one single mechanism is not sufficient to describe the full range of binder degradation. To clarify these different effects, we first simulate weight loss, *i.e.*, TGA, data using a first order kinetic mechanism, $f(\alpha)=1-\alpha$, for the full range of conversion with values of $A=1.3\times10^3$ s^{-1} and $E=57$ kJ/mol. Figure 1 shows that the simulated TGA profiles obtained in this way exhibit the classic shape of first order decomposition.

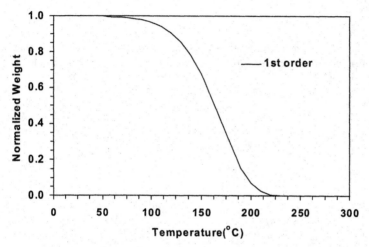

Fig. 1 Simulated weight loss data with a first order mechanism with $A=1.3\times10^3$ s^{-1} and E 57 kJ/mol for a heating rate $\beta=0.5°C/min$.

The weight loss profile in Fig. 1 was then analyzed with Eq. 7 for all of the decomposition mechanisms in Table I. The values of A and E reported in Table II are obtained for a truncated range of conversion; over this truncated range, the integral method of analysis leads to linear behavior with regression coefficients near unity, as is indicted in Table II. For each assumed kinetic mechanism $f(\alpha)$, very different values of A and E are obtained. The values of A and E obtained for each mechanism also vary in the same way, in that large values of E correspond to large values of A, a so-called compensation effect.

We next simulated weight loss profiles for all of the kinetic mechanisms in Table I using the corresponding values of A and E from Table II. Figure 2 shows that all of the mechanisms describe reasonably accurately the truncated range of conversion of binder degradation indicted in Table II. At high levels of conversion, however, the simulated profiles diverge from the simulated kinetics, with the poorest level of agreement being obtained from the zero- and second-order models.

Table II Preexponential factor and activation energy determined from simulated first-order binder decomposition with $A=1.29\times10^3$ s^{-1} and $E=57$ kJ/mol. The regression coefficient for the goodness of fit for a truncated range of conversion (1-α) is also indicated.

Kinetic Mechanism	E (kJ/mol)	A (s^{-1})	r^2	1-α
0 order	54.4	5.65×10^2	0.999	1.00~0.41
1st order	57.0	1.29×10^3	1.000	1.00~0.00
2nd order	58.6	2.78×10^3	0.999	1.00~0.55
Diffusion A	115.2	6.16×10^9	0.999	1.00~0.27
Diffusion B	117.3	3.00×10^9	0.999	1.00~0.27

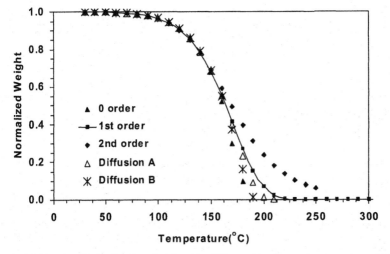

Fig. 2 Simulated kinetics (symbols) with different kinetic mechanisms from Table I with the corresponding values of A and E from Table II determined for a truncated range of conversion. The original first order data is shown by the solid line.

The simulated kinetics in Fig. 1 were next analyzed over the full range of weight loss for α between 0 and 1. For this case, the values of A and E in Table III are different from what is reported in Table II for the truncated range of conversion. The corresponding regression coefficients are also contained in Table III; in general, the regression coefficients are still high but not as high as for when a truncated range of conversion is analyzed. Figure 3 shows the simulated kinetics for the kinetic parameters in Table III with the corresponding mechanisms from Table I. The level of agreement is now poorer in the middle range of conversion but both tails of the weight loss data are reasonably well described. The poorest level of agreement is now seen for the second order and Diffusion B mechanisms.

Table III Preexponential factor and activation energy determined from simulated first-order binder decomposition with $A=1.29\times10^3$ s^{-1} and E=57 kJ/mol. The regression coefficient for the goodness of fit for the full range of conversion (1-α) is also indicated.

Kinetic Mechanism	E (kJ/mol)	A (s^{-1})	r^2	1-α
0 order	47.1	1.50×10^2	0.972	1.0~0.0
1st order	57.0	1.29×10^3	1.000	1.0~0.0
2nd order	70.3	1.95×10^5	0.958	1.0~0.0
Diffusion A	104.9	1.70×10^8	0.996	1.0~0.0
Diffusion B	111.5	3.96×10^8	0.985	1.0~0.0

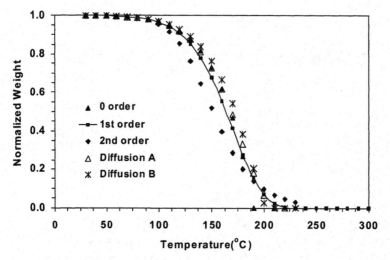

Fig. 3 Simulated kinetics (symbols) with different kinetic mechanisms from Table I with the corresponding values of A and E from Table III determined for the full range of conversion. The original first order data is shown by the solid line.

The accuracy seen in Figs. 2 and 3 for the weight loss profiles also holds when the data are initially simulated with other mechanisms or with other values of A and E. In general, then, the values of A and E obtained from the analysis of TGA data depend on the form of the mechanism assumed in the analysis and on the range of weight loss data analyzed. This exercise indicates that unless the kinetic mechanism is definitively known, which is rarely the case, then it may not be possible to unambiguously obtain a set of unique kinetic constants. The behavior observed in Figs. 2 and 3 further indicate that although it is not difficult to represent TGA data when the mechanism is unknown, very different kinetic parameters will arise which will then ultimately be used to predict heating profiles for binder removal, as indicated by Eqs. 5 and 9.

Ultimately, however, we are interested in showing how TGA data for real binder systems can be analyzed. Figure 4 shows TGA weight loss data we have obtained for the decomposition in air of poly(vinyl)butyral plasticized with dibutylphthalate. At least three main regions of binder decomposition are evident, and the kinetic parameters corresponding to these three regions are contained in Table IV. For each region, very different values of A and E arise, with high values of the regression coefficients for each region.

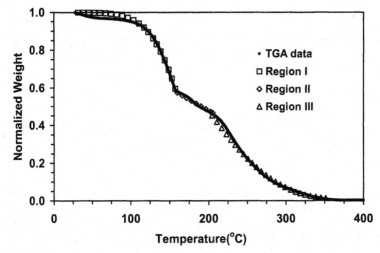

Fig. 4 TGA data (solid line) for the decomposition in air of poly(vinyl)butyral and dibutyl phthalate for a heating rate of 0.5°C/min. The simulated kinetics (symbols) for the three regions of binder decomposition are also shown.

Table IV Preexponential factor and activation energy determined for the decomposition in air of poly(vinyl)butyral and dibutyl phthalate with a first-order mechanism for a heating rate of 0.5°C.

Region	E (kJ/mol)	A (s⁻¹)	r^2	1-α	Δα
I	57	1.29×10^3	0.99	0.94~0.58	0.38
II	7	2.70×10^{-4}	0.98	0.58~0.42	0.17
III	21	2.20×10^{-2}	0.99	0.42~0	0.45
Average	32	6.75×10^{-1}	NA	1.0~0	NA

Figure 4 also shows the TGA profiles simulated with these values, and reasonable agreement is obtained over the full range of conversion. To obtain such good agreement, however, the kinetics need to be simulated in a piecewise fashion, corresponding to the three ranges of conversion. This can be represented mathematically for a first order process as

$$r = A\exp(-\frac{E}{RT})\frac{\varepsilon_b}{\varepsilon_{b,o}} \quad \text{with} \begin{cases} A = 1.3\times10^3 \ s^{-1}; E = 57\,kJ\,/\,mol\,for \quad 0 \le \alpha < 0.42 \\ A = 2.7\times10^{-4}\,s^{-1}; E = 7\,kJ\,/\,mol\,for \ 0.42 \le \alpha < 0.58 \\ A = 2.2\times10^{-2}\,s^{-1}; E = 21\,kJ\,/\,mol\,for \ 0.58 \le \alpha < 1 \end{cases} \quad 10$$

This lack of a continuous kinetic representation, however, will lead to difficulties in determining a continuous and smooth heating profile for binder removal. To circumvent this difficulty, the kinetic constants in Table IV for the three regions of decomposition in Fig. 4 can be averaged to obtain a single set of A and E. To perform this averaging, a weighted arithmetic average based on the range of conversion, $\Delta\alpha$, is taken for the activation energy and a weighted geometric average based on $\Delta\alpha$ is taken for the preexponential factor, as indicted below for n separate regions of weight loss:

$$\overline{E} = \sum_{i=1}^{n} \Delta\alpha_i E_i \qquad\qquad 11$$

$$\overline{A} = \sqrt[n]{\prod_{i}^{n} \Delta\alpha_i A_i} \qquad\qquad 12$$

These weighted average values are also listed in Table IV. Figure 5 shows that although the TGA data simulated with these average values do not lead to as an accurate representation of the kinetic data as using a piecewise continuous function, the simulated kinetics do capture, to some extent, the temperature range over which binder degradation occurs. Although clearly not as accurate as the piecewise continuous model, such an averaging method is advantageous in light of being a continuous function and may be sufficient in light of the overall uncertainty associated with a number of other parameters that appear in the model. In addition, the average kinetic constants do provide a reasonable representation of the TGA data in the important early part of the heating cycle, when the green body is highly loaded with binder.

CONCLUSIONS

Determination of a kinetic mechanism and the values of the activation energy and preexponential factor for binder decomposition from green ceramic bodies is an important aspect necessary to determine the minimum cycle times for binder removal. In this work, we illustrate with both simulated and real weight loss data how the kinetics can be determined. Depending on the assumed mechanism and the range of conversion evaluated, different values of the activation energy and preexponential factor can be determined. These values can then be used to obtain a reasonably accurate representation of the binder decomposition behavior. We also show how TGA data for a real binder system can be analyzed as either a piecewise continuous function or by using a single set of averaged values for the activation energy and preexponential factor.

ACKNOWLEDGMENT

This project was funded by Honeywell Federal Manufacturing & Technologies, which is operated for the United States Department of Energy, National Nuclear Security Agency, under the contract No. DE-AC04-01AL66850.

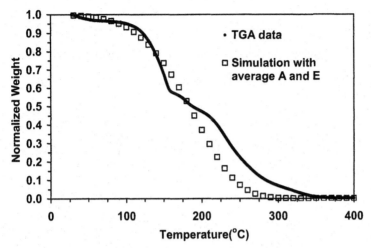

Fig. 5 TGA data (solid line) for the decomposition in air of poly(vinyl)butyral and dibutyl phthalate with a first-order mechanism for a heating rate of 0.5°C. The simulated kinetics (symbols) with the average values of A and E are also shown.

NOMENCLATURE

L_i, i=x,y,z dimensions of the body
R, gas constant
r, rate of binder decomposition
A, preexponential factor
E, activation energy
t, time
t^*, minimum cycle time for binder removal
T_o, initial temperature
T_s, starting temperature of binder removal cycle
P_o, initial ambient pressure
P_t, threshold pressure in the center of the green ceramic body
α, binder conversion
ε_b, volume fraction of binder
$\varepsilon_{b,o}$, initial volume fraction of binder
ρ_b, binder density
ρ_o, initial gas density at T_o and P_o
M, average molecular weight of gas products
R, universal gas constant
P_t, threshold pressure in the center of the body corresponding to failure
μ, gas viscosity
κ, permeability
S, surface area per unit volume
k, Kozeny-Carman parameter

REFERENCES

1. R. M. German, "Theory of Thermal Debinding," *Int. J. Powder Metall.,* **23**, 237-245 (1987).
2. J. A. Lewis, "Binder Removal From Ceramics," *Annual Rev. Mater. Sci.,* **27**, 147-173 (1997).
3. P. Calvert and M. Cima, "Theoretical Models for Binder Burnout," *J. Am. Ceram. Soc.,* **73**, 575-579 (1990).
4. G. Y. Stangle, and I. A. Aksay, "Simultaneous Momentum, Heat and Mass Transfer With Chemical Reaction in A Disordered Porous Medium: Application to Binder Removal From A Ceramic Green Body," *Chem. Eng. Sci.,* **45**, 1719-1731 (1990).
5. D.-S. Tsai, "Pressure Buildup and Internal Stresses During Binder Burnout: Numerical Analysis", *AIChE J.,* **37**, 547-554 (1991).
6. S. A. Matar, M. J. Edirisinghe, J. R. G. Evans, and E. H. Twizell, "Effect of Porosity Development on the Removal of Organic Vehicle from Ceramic or Metal Moldings," *J. Mater. Res.,* **8**, 617-625 (1993).
7. J. H. Song, M. J. Edirisinghe, J. R. G. Evans, and E. H. Twizell, "Modeling the Effect of Gas Transport on the Formation of Defects during Thermolysis of Powder Moldings," *J. Mater. Res.,* **11**, 830-840 (1996).
8. A. C. West and S. J. Lombardo, "The Role of Thermal and Transport Properties on The Binder Burnout of Injection Molded Ceramic Components," *Chem. Eng. J.,* **71**, 243-252 (1998).
9. S. J. Lombardo and Z. C. Feng, "Pressure Distribution during Binder Burnout in Three-Dimensional Porous Ceramic Bodies with Anisotropic Permeability," *J. Mat. Res.,* **17**, 1434-1440 (2002).
10. Z. C. Feng, B. He, and S. J. Lombardo, "Stress Distribution in Porous Ceramic Bodies During Binder Burnout," *J. of Appl. Mech.,* **69**, 497-501 (2002).
11. K. Feng and S. J. Lombardo, "Modeling of the Pressure Distribution in Three-Dimensional Porous Green Bodies during Binder Removal," *J. Am. Ceram. Soc.* **86**, 234-240 (2003).
12. S. J. Lombardo and Z. C. Feng, "Determination of the Minimum Time for Binder Removal and Optimum Geometry for Three-Dimensional Porous Green Bodies," *J. Am.Ceram.Soc.,* **6** (2003) 2087-2092.
13. S. J. Lombardo and Z. C. Feng, "Analytic Method for the Minimum Time for Binder Removal from Three-Dimensional Porous Green Bodies," *J. of Mat. Res.,* **18** (2003) 2717-2723.
14. C. David, Ch. 1 in *Comprehensive Chemical Kinetics-Degradation of Polymers,* edited by C H. Gamford and C.F.H. Tipper, Vol. 14, Elsevier, New York, 1975.
15. H. H. G. Jellinek, "Degradation and Depolymerization Kinetics," pp. 1-37 in *Aspects of Degradation and Stabilization of Polymers,* edited by H. H. G. Jellinek (Elsevier, NY 1978).
16. R. V. Shende and S. J. Lombardo, "Determination of Binder Decomposition Kinetics for Specifying Heating Parameters in Binder Burnout Cycles," *J. Am. Ceram. Soc.,* **85**, 780-786 (2002).
17. L. C.-K. Liau, B. Peters, D. S. Krueger, A. Gordon, D. S. Viswanath, and S. J. Lombardo, "Role of Length Scale on Pressure Increase and Yield of Poly(vinyl butyral)-Barium

Titanate-Platinum Multilayer Ceramic Capacitors During Binder Burnout," *J. Am. Ceram. Soc.*, **83**, 2645-2653 (2000).

18. S. J. Lombardo and R. V. Shende, "Determination of Polymer Decomposition Kinetics to Specify Ramps and Holds for Binder Burnout Cycles for Multilayer Ceramic Capacitors," pp. 23-31 in *Advances in Dielectric Materials and Multilayer Electronic Devices, American Ceramic Society*, edited by K. M. Nair, A. S. Bhalla, and S.-I. Hirano, in Ceramic Transactions, Vol. 131, Westerville, OH, 2002.
19. T. V. Lee and S. R. Beck, "A New Integral Approximation Formula for Kinetic Analysis of Nonisothermal TGA Data," *AIChE J.*, **30**, 517-519 (1984).

Computational Modeling and Analysis

STRESS-DEFORMATION ANALYSIS BY FEM-BASED COMPUTATIONAL MODELING OF FRACTURE MECHANICS OF CARBON HARD INCLUSIONS AND REINFORCED COMPOSITE MATERIALS

Maksim V. Kireitseu and David Hui
Composite Nano/Materials Research Center
University of New Orleans,
New Orleans, LA 70148-2220, USA
E-mail: m.kireitseu@tridentworld.com

Liya Bochkareva
United Institute of Informatics Problems NAS,
Filatova str., 7 - 28, Minsk 220026,
Belarus

Sergey Eremeev and Igor Nedavniy
Institute of Strength Physics and Materials Science,
Siberian Branch of Russian Academy of Sciences,
2/1 Academicheskii Pr., Tomsk 634021, Russia

ABSTRACT

This paper reports a study of the mechanical behavior of reinforced composite nanomaterials and carbon-based hard nanoparticles/fibers such as nanotubes (MWCNTs). The nested individual layers of carbon nanoparticles/fibers are treated as single-walled frame-like structures and simulated by the molecular structural mechanics method. The volumes containing significant elements of internal submicrocrystalline and nano structure of material are investigated by numerical methods of continuum mechanics. The purpose of the research is numerical modeling of mesovolumes of such materials.

INTRODUCTION

Since the discovery of multi-walled carbon-based materials (nanotubes and diamonds) (MWCNT), NASA centers and researchers worldwide have engaged in fundamental studies of this novel material and have investigated the potential of its applied engineering and technological applications, including carbon nanotube and carbon nanoparticles (diamonds) based composites. The unique characteristics of carbon-based materials (nanotubes and diamonds), including diamonds nanoparticles, such as high stiffness, high aspect ratio and low density, are particularly desirable for this new generation of composites for aerospace engineering, offshore mechanics and marine technologies. In order to fully explore the potential of carbon-based materials (nanotubes and diamonds) for application in various composites, a thorough understanding of the mechanical properties of carbon-based materials (nanotubes and diamonds) and diamonds is necessary.

The advancement of science and technology has evolved into the era of nanotechnology. Because 3D computer simulations based on reasonable physical models cannot only highlight the molecular features of nanomaterials for theoreticians but also provide guidance and interpretations for experimentalists. It is still an ongoing and challenging process to identify effective and efficient computational methods with respect to specific nanomaterials.

Among the many nanostructured materials, carbon-based nanoparticles (diamonds and nanotubes) have attracted considerable attention. They can be produced by an array of techniques, such as arc discharge, laser ablation and chemical vapor deposition. From the viewpoint of atomic arrangement, carbon-based materials (nanotubes and diamonds) can be visualized as cylinders that rolled from sheets of graphite. Whereas diamonds may have various shapes: pyramid, sphere and fullerene-like. One shape may be reconstructed to another. Due to their unique molecular structure, carbon nano-tubes possess excellent mechanical properties like high stiffness and strength which makes them attractive as potential reinforcement of materials [1, 3].

A single-walled nanotube is a hollow structure formed by covalently bonded carbon atoms. It can be imagined as a rectangular graphene sheet rolled from one side of its longest edge to form a cylindrical tube. Hemispherical caps seal both ends of the tube as shown in Fig. 1, 2. For multi-walled nanotubes, a number of grapheme layers are co-axially rolled together to form a cylindrical tube.

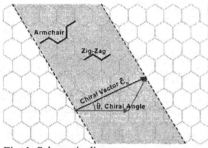

Fig. 1. Schematic diagram
of a hexagonal graphene sheet

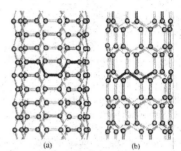

Fig. 2. Schematic diagram of (a) an armchair
and (b) a zigzag nanotube

The adequate description and numerical simulation of features of deformation and fracture of materials at various kind of loading are of great importance to the problem of design of advanced reinforced materials. It is well known from experiments [8, 9] that plastic deformation is generated irregularly and its development is accompanied with formation of various regions of localized strain of different scales at pre-fracture stage. At meso and macro levels there are shear bands, thinning regions, Lüders bands, neck formation, etc. One subject of this paper is the problem of computer simulation of such phenomena.

The approach of mesomechanics of solids, the conception of scale and structural levels of deformation and fracture are used here [9, 10]. Just on, mesoscale there are basic features of plastic deformation not investigated yet. By mesovolume, we mean a volume of a material in which we obviously deal with heterogeneities of internal structure having various physical-mechanical properties. In such mesovolumes of material, the medium is non-homogeneous. It has a structure, which evolve in course of loading. Some heterogeneities of the structure can be taken into account explicitly and other non-explicitly, using various complex continuous medium model. Heterogeneity of internal structure of a material causes heterogeneity of the stress-strain state in material under loading. The main features of stress-strain state observed in experiment should be received in simulations.

Here finite difference analysis of model mesovolumes of different materials is carried out. We use simple isotropic elastic plastic model and pay especial attention on inhomogeneity of materials at the mesolevel. Though quasi-static loading is considered, dynamic numerical method is applied. In this paper, we extend the theory of classical structural mechanics into the modeling of carbon-based materials (nanotubes and diamonds).

Thus, it is logical to simulate the deformation of a nanotube based on the method of classical structural mechanics. In following sections, we first establish the bases of this concept and then demonstrate the approach by a few computational examples.

MULTISCALE MODELING OF MATERIALS: OVERVIEW

We have made an attempt to provide a comprehensive overview and analysis of fundamental techniques and definitions of the current status of multiscale simulations methods and their applications to carbon-based nanocomposites. Fundamentally, we divide the methods of multiscale problems into sequential and concurrent.

The sequential techniques of multiscale modeling of nanomaterials are in general more efficient computationally, but they depend on a priori knowledge of physical quantities of interest, such as the γ-surface in the P-N model, the free-energies in the phase-field model, and atomistic local laws for mesoscopic DD simulations. The relevance of these quantities to the coarse-grained models needs to be care- fully examined before the application of the methods. Furthermore, these approaches should only be pursued when phenomenological theories (such as the P-N model or the phase-field model) are well established; therefore the methods are restricted in their range of application.

In particular, these phenomenological models are often associated with the assumption of locality (both in space and time). The example of a local approximation in the phase-field model is embodied in Eq. (*), which assumes that part of the energetics of an inhomogeneous system can be written in terms of quantities obtained for homogeneous systems [1]. Similarly, in the P-N model, the γ-energy is assumed to be constant within _x distance in order to evaluate the total misfit energy. The static approximation (locality in time) for dynamical properties is also widely used in phenomenological models. The coupling between different scales in a sequential approach is usually implicit. A successful sequential simulation depends equally on the reliability of the phenomenological model and the accuracy of the relevant parameters entering the model.

The concurrent multiscale approaches are much more complicated and computationally demanding, but they do not require a priori knowledge of physical quantities supplied from distinct, lower scale simulations. Furthermore, concurrent approaches do not depend on any phenomenological models; therefore they are of more general applicability. Although concurrent approaches are more desirable and appealing, the actual problem to be attacked must be carefully posed in order to make the method practical. The problems that may arise in a concurrent approach are usually associated with the partition of domains in the system. For example, one needs to dynamically track the domain boundaries in the MAAD simulations and to adapt the FE meshes in the quasi-continuum simulations, both of which require additional care and computational resources.

For a structural transformation, the total free energy can be written as:

$$F_{tot} = F_{bulk} + F_{inter} + F_{elast} \qquad (*)$$

where F_{bulk} is the bulk free energy, F_{inter} is the interfacial free energy, and F_{elast} is the coherency elastic strain energy arising from the lattice accommodation along the coherent interfaces in a microstructure. For a microstructure described by a composition field c and a set of structural order-parameters, $\eta_1, ..., \eta_p$, the first two terms of Eq. (*) are given by

$$F_{bulk} + F_{inter} = \int_V \{ f[c(r), \eta(r) + \frac{\alpha}{2} |\nabla c(r)|^2 + \frac{1}{2} \sum_p \beta_{ij}(p) \nabla_i \eta_p(r) \nabla_j \eta_p(r) \} dV \qquad (**)$$

where f(c, ηp) is the local free energy density [6] and α and βij (p) are the gradient energy coefficients which control the width of the diffuse interface. The elastic strain energy is obtained from elasticity theory using the homogeneous modulus approximation [7].

Within this approach, the equilibrium structure of a dislocation is obtained by minimizing the dislocation energy functional with respect to the dislocation misfit density [8].

$$U_{dist} = U_{elastic} + U_{misfit} + U_{stress} + Kb^2 \ln L \qquad (***)$$

Here, $\rho(1)_i$, $\rho(2)_i$ and $\rho(3)_i$ are the edge, vertical and screw components of the general interplanar misfit density at the i-[th] nodal point, and $\gamma_3(f_i)$ is the corresponding three-dimensional γ-surface. The components of the applied stress interacting with the $\rho(1)_i$, $\rho(2)_i$ and $\rho(3)_i$, are $\tau(1) = \sigma_{21}$, $\tau(2) = \sigma_{22}$ and $\tau(3) = \sigma_{23}$, respectively. K, Ke and Ks are the pre-logarithmic elastic energy factors. The dislocation density at the i-th nodal point is $\rho_i = (f_i - f_i - 1) / (x_i - x_i - 1)$ and χ_{ij} is the elastic energy kernel [8].

More importantly, in contrast to a sequential method, a "good" hand-shaking in a concurrent approach between different domains is both challenging and critical. Although some interesting ideas have been proposed to remedy the problems of coupling between different domains, such as the reflection of phonons at the domain interface [1-3], there is no general consensus on what a proper coupling of domains is. A general criterion that measures the quality of hand-shaking between domains would therefore be desirable. There is plenty of room for innovative research on the issue of domain coupling. General mathematical formulations of multiscale problems, including error estimation, may turn out to be very useful for practical simulations [4, 5].

In our view, a successful concurrent approach usually has to satisfy three conditions:

(1) Solve the coupled problem (Hamiltonian) accurately and efficiently by using ideas such as coarse-graining, "bonding energy partition" or "embedding potential";

(2) The separate models employed in different domains of the system ought to be compatible, i.e., the physical description of the system due to the distinct models should be as close as possible;

(3) At each level, the individual model should provide a good description of its assigned domain.

We wish to emphasize the importance of the second condition which is not in general well recognized. The first condition usually guarantees a "smooth" hand-shaking across two domains (e.g., the electron density distribution or the displacement field varies smoothly across the interface), but non-physical charge transfer and/or atomic relaxation at the interface could occur if the second condition is not satisfied. Therefore a "smooth" hand-shaking does not constitute a "good" hand-shaking, and a successful concurrent approach relies on hand-shakings that are both mathematically accurate and physically consistent.

In this overview, we cannot mention in details a number of approaches that strive to extend the temporal scale in the modeling and simulation of material properties. We categorized these approaches to methods for accelerating the dynamics, methods for finding transition paths between equilibrium structures, and methods for escaping free-energy minima. Although these approaches represent very significant developments in the field, the problem of linking the time scale of atomic motion and vibrations (of order femtoseconds) to scales where interesting physical phenomena are typically studied (microseconds and beyond) is still wide open in many respects.

We propose the following computational technique of finite element method modeling of nanostructures (carbon-based structures C):

1. Force field model of carbon-based structures;
2. Finite element method modeling, where definition of finite element is determined;
3. Equivalency of Molecular dynamics and Finite element (FEM) methods should be shown;
4. Technique of mathematical FEM-based modeling of nanostructure (graphite – 2D, nanotube – 3D);
5. Development of model of graphite (2D model) to demonstrate a problem of mathematical description of 2D and 3D nanostructures;
6. Calculations of mechanical properties of carbon-based nanostructures by finite element method (Young modulus, Poisson ratio, shear stress, stress and deformation).

Because of the tremendous and continuing progress in multiscale strategies, this review and proposed FEM-based technique is by no means exhaustive. We hope that we have conveyed the message that multiscale modeling is a truly vibrant enterprise of multi-disciplinary nature. It combines the skills of physicists, materials scientists, chemists, mechanical and chemical engineers, applied mathematicians and computer scientists. The marriage of disciplines and the concomitant dissolution of traditional barriers between them represent the true power and embody the great promise of multiscale approaches for enhancing our understanding of, and our ability to control complex physical phenomena.

STRUCTURAL MECHANIC APPROACH

The potential energy of molecular systems like carbon nanotubes can be calculated with the force field method. We apply the DREIDING approach which is described by [4], among others. Neglecting inversion and non-bonded interactions, the potential energy of this generic force field is of the form

$$E = E_B + E_A + E_T \qquad (1)$$

Fig. 3 illustrates the bond stretch energies. The distance between two atoms is R_e=1.39 Å at the equilibrium state. For the bond order n=1.5, the bond energies are given to k_e=1050 kcal/(mol Å²) and D_e=105 kcal/mol.

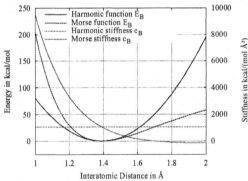

Fig. 3 Bond stretch energies

The bond between two atoms I and J can be described by the harmonic potential

$$E_B = \tfrac{1}{2} k_e \left(R - R_e \right)^2 \qquad (2)$$

which leads to the spring constant

$$c_B = \frac{d^2 E_B}{dR^2} = k_e \tag{3}$$

or by the Morse function

$$E_B = D_e \left[e^{-an(R-R_e)} - 1 \right]^2 \tag{4}$$

which yields the nonlinear spring coefficient

$$c_B = \frac{d^2 E_B}{dR^2} = 4D_e (\alpha n)^2 \left[e^{-2\alpha n(R-R_e)} - \tfrac{1}{2} e^{-\alpha n(R-R_e)} \right] \tag{5}.$$

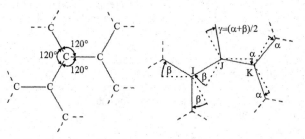

a.) Undeformed configuration   b.) Deformed configuration

Fig. 4 Bond angles

The second term of DREIDING

$$E_A = E_{IJK} = \tfrac{1}{2} K_{IJK} \left[\theta_{IJK} - \theta_J^0 \right]^2 \tag{6}$$

describes the variation of the potential energy caused by change of bond angles, see Fig. 4.

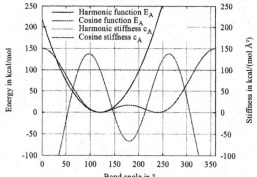

Fig. 5. Bond angle energies

This energy function leads to the spring constant:

$$c_A = \frac{d^2 E_A}{d\theta_{IJK}^2} = K_{IJK} \tag{7}$$

Alternatively, DREIDING provides a cosine form

$$E_A = E_{IJK} = \frac{1}{2} C_{IJK} \left[\cos\theta_{IJK} - \cos\theta_J^0 \right]^2 \tag{8}$$

with the nonlinear spring coefficient

$$c_A = \frac{d^2 E_A}{d\theta_{IJK}^2} = C_{IJK} \left[\cos\theta_J^0 \cos\theta_{IJK} - \cos\left(2\theta_{IJK} \right) \right] \tag{9}$$

With $\theta_j^0 = 120°$ and $K_{IJK} = 100$ kcal/(mol rad²), we obtain the bond angle energies given in Fig. 4. The torsion energy

$$E_T = E_{IJKL} = \frac{1}{2} V_{JK} \left\{ 1 - \cos\left[n_{JK} \left(\varphi - \varphi_{JK}^0 \right) \right] \right\} \tag{10}$$

depends on the dihedral angle φ between the planes IJK and JKL. The torsional spring constant is

$$c_T = \frac{d^2 E_T}{d\varphi^2} = \frac{1}{2} V_{JK} n_{JK}^2 \cos\left[n_{JK} \left(\varphi - \varphi_{JK}^0 \right) \right] \tag{11}$$

The equilibrium angle is $\varphi_{jk}^0 = 180°$, the periodicity $n_{JK} = 2$ and the torsion energy $V_{JK} = 25$ kcal/(mol rad²).

FINITE ELEMENT MODELLING

It has to be considered that the bond angle energy $E_A = E_{IJK}$ only depends on the angle between the bonds IJ and JK. As illustrated in Fig. 4b, the adjacent angles are equal to α and β if E_{IJK} is the only energy. Frame elements (Fig. 6a) force the adjacent bonds to rotate, but due to their bending resistance, these angles are unequal to α and β. Since this could only be prevented by additional constraints, we prefer a finite element model using spring elements (Fig. 6b).

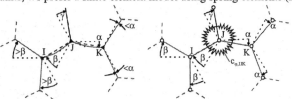

a.) Beam elements   b.) Non-linear spring elements
Fig. 6 Simple and improved beam models

In order to be able to compute large structures containing several nanotubes, the number of degrees of freedom of the overall finite element model has to be reduced. Shell elements could be used for that purpose, see e.g. [2,5,6,7]. Fig. 7 shows the differences between these finite element models.

Because 3D computer simulations based on reasonable physical models cannot only highlight the molecular features of nanomaterials for theoreticians but also provide guidance and interpretations for experimentalists. It is still an ongoing and challenging process to identify effective and efficient computational methods with respect to specific nanomaterials.

However, it's true that carbon-based nanomaterials and nanocomposites have been the subject of enormous interest, but development of nanotechnology has been spurred by refinement of IT engineering tools to virtually design the nanoworld, such as more sophisticated information technologies, software and computers to be used for nanoengineering [1]. The principal objective of the paper was to demonstrate an application of modern software engineering tools for modeling virtual reality and molecular dynamics of novel nanocomposites.

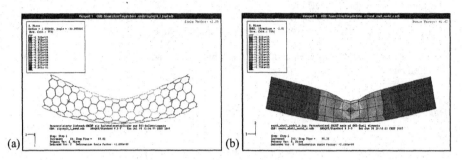

Fig. 7. Beam element model (a) and Shell element model (b)

COMPUTER TOOLS FOR 3D MODELING

Our joint group is exploring the use of spatially immersive virtual reality systems and modern IT technologies (e.g., Visual Studio Microsoft.Net, Active X and JAVA2 applets) for interactive modeling and visualization of nanotechnology relevant systems. A computational scheme and software, which utilizes neural networks and/or Microsoft.Net technique, was developed to predict properties of nano-structured materials and optimization and control of nano-devices (fig. 8).

The 3D world of different topologies and shapes

Diamonds and fullerene-like nanostructures

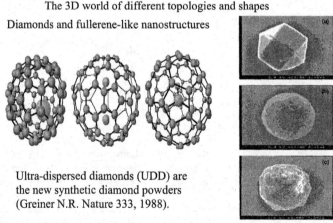

Ultra-dispersed diamonds (UDD) are the new synthetic diamond powders (Greiner N.R. Nature 333, 1988).

Fig. 8. Carbon-based diamonds nanostructures (virtual and real)

We studied ultra dispersed diamond nanoparticles having different size and shape. The nanoparticles could be pyramid, sphere-like and fullerene-like shape. Its size might be up to 100 nm and higher, but usually their diameter is between 5 and 15 nm. Computer-vision-assisted techniques enable semi-automatic intermittent 3-D graphic model update to match the simulated virtual reality of nanomaterials with their actual video images. In result (fig. 8) we may easily have the 3D virtual world of nanostructures and nanocomposites of different shapes.

Developed software and (info/analytical tools) IT tools for modelling of virtual reality of nanostructures and nanocomposites differs from others either on the market or under development as follows.

1. Developed a computational scheme and software utilizes neural networks and/or Microsoft.Net technique that were developed just a year ago by Microsoft Inc. as an advanced analogue of C++ or C Sharp programming languages. Now it differs by developed three-level architecture of information systems and tools for virtual nanoengineering. Main concept of the developed tools includes additional novel level of persistent nanoobjects modelled in addition to known Miscosoft.NET platform.

2. Developed an unique set of simple algorithms to encode the structure and composition of the nanomaterial directly into numerical vectors neural network modules can correlate these numeric inputs with a set of desired properties. The interactive system employs a reactive diamond potential coupled to a virtual three-degree-of-freedom hap tic (force-feedback) arm and a real-time graphic display through Internet or intranet.

3. The software and nanoengineering tools to be announced are based on three key parts: server of applications and database of nanoobjects, tools for engineering and modelling of 3D virtual reality and tools for transferring through internet-based solutions and cable networks, utilities for administration. The tools are easily integrated with known platforms and databases such as Rational Rose, MS SQL, MS Access, Oracle DB and others. The nanoengineering tools also provide independence between logical structure of modelled nanomaterials and its physical core. It made universal engineering logic for various databases, software shells and interfaces.

A software system provides for semi-automated calibration of three-dimensional computer-generated images (e.g., "virtual reality" images) from digitized images of nanomaterials. Three-dimensional graphic models are intermittently updated through this virtual-reality calibration, which determines the scanning microscope and camera calibration parameters and nanoobject locations semi-automatically by using model-based, edge-matching computer-vision algorithms. Spatially immersive 3D display (SID) devices surround the user in real space with a 3D computer-generated visual and audio scene that is responsive to the user's point of view, orientation, and action. SIDs typically are based on multiple large-format stereo projection systems and can provide 3D views to multiple users.

4. Developed problem tracking system to be used during modeling of virtual reality of nanostructures and nanocomposites including the following stages: detecting, registration, processing, localization of occurred errors or lacks, testing, modification and proposing a few possible solutions, delivery of solutions to a client or an engineer. Detected problems are classified for internal using by engineers.

VIRTUAL NANOENGINEERING

The user can easily attach his hand to an atom or molecule and maneuver it in 3D three-dimensional space. Resulting picture (fig. 9-11) of nano simulation shows respectively, for example, crystallise lattice of aluminium matrix in fig. 9, diamond nanoparticles of chosen shape in fig. 10 (pyramid-like, sphere or fullerene-like shapes) and nanocomposites represented as aluminium matrix with two introduced pyramid-like diamond nanoparticles (green-atoms in fig. 11). The positions of all atoms in the atomic lattices are also calculated and stored in data file. The data are used for further calculations of mechanical properties by the number of techniques (FEM, FEM, MD etc).

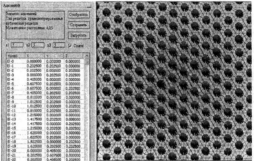

Fig. 9. Atomic lattice of Aluminium matrix.

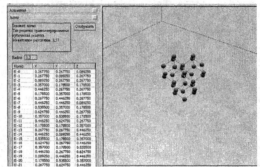

Fig. 10. Atomic lattice of diamond nanoparticle.

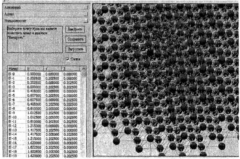

Fig. 11. Reinforced Nanocomposite: Aluminium matrix – diamond nanoparticles

Valuable solution and tool for modelling of nanostructures and nanomaterials provide maximum performance and quality of obtained 3D video images. The reactive potential and mechanical properties are calculated from the resultant forces and motions of all atoms and nanoparticles, including changes in bonding topology. The forces on the atoms being manipulated are continuously fed back to the user through the haptic interface, while the newly calculated positions of all atoms and nanoparticles are continuously updated on the 3D graphics display.

We also believe that nanotechnology CAD environments and SIDs offer significant advantages relative to traditional displays or head mounted displays for collaboratively designing and exploring engineered nanostructures. Virtual database of various nanostructures and their computer-based calculations will introduce outstanding possibilities to researchers in this field.

MODEL AND METHOD OF 2D SIMULATION

The local field solution shows large gradients in deformations and in stresses due to the rigid inclusions. These gradients are difficult to solve using FEM techniques - a very fine mesh is required. More convenient for handling such large gradients are reciprocity based Boundary Element (BE) techniques based on non-singular integral equation formulations. In such formulation the source points are located outside the domain. Moreover, the boundary displacements and boundary tractions are modeled in different ways, which enables to reduce the problem considerably and to obtain very good accuracy of the solution. The stress fields are obtained using low (second order) Trefftz polynomial interpolators. The solution is shown for 2D problems, but the extension to 3D is straightforward.

The Lagrange approach is used. 2D problems under plane strain and plane stress conditions were considered. The usual set of equations of dynamics of deformable solids for the case of two dimensional plane elastic-plastic flow with von Mises yield criterion was used [11, 12]. The basic equations are the following. Equations for strain rates:

$$\dot{\varepsilon}_{xx} = \frac{\partial v_x}{\partial x}, \; \dot{\varepsilon}_{yy} = \frac{\partial v_y}{\partial y}, \; \dot{\varepsilon}_{xy} = \frac{1}{2}\left(\frac{\partial v_y}{\partial x} + \frac{\partial v_x}{\partial y}\right), \; \dot{\omega}_z = -\dot{\omega}_{xy} = \frac{1}{2}\left(\frac{\partial v_y}{\partial x} - \frac{\partial v_x}{\partial y}\right) \quad (12)$$

$$\dot{\varepsilon}_{zz} = \frac{\dot{h}}{h} \text{ - for plane stress state, } \varepsilon_{zz} = 0 \text{ - for plane strain state.}$$

Equations of motion in plane x-y coordinates can be written as follows

$$\rho\frac{\partial v_x}{\partial t} = \frac{\partial \sigma_{xx}}{\partial x} + \frac{\partial \sigma_{xy}}{\partial y}, \; \rho\frac{\partial v_y}{\partial t} = \frac{\partial \sigma_{xy}}{\partial x} + \frac{\partial \sigma_{yy}}{\partial y} \quad (13)$$

Equation of continuity:

$$\frac{\dot{V}}{V} = \dot{\varepsilon}_{xx} + \dot{\varepsilon}_{yy} + \dot{\varepsilon}_{zz}.$$

Equations of state:

$$\sigma_{ij} = -P\delta_{ij} + s_{ij}, \; \dot{\varepsilon}_{ij} = \dot{\varepsilon}_{ij}^e + \dot{\varepsilon}_{ij}^p, \; \dot{\varepsilon}_{ij}^p = \dot{\lambda} s_{ij},$$

$$\dot{P} = -K\frac{\dot{V}}{V} \text{ - for stresses lower than 1 GPa and barotropic model of medium,}$$

$$\frac{Ds_{xx}}{Dt} = 2\mu\left(\dot{\varepsilon}_{xx} - \frac{1}{3}\frac{\dot{V}}{V}\right) - 2\mu\lambda s_{xx}, \qquad \frac{Ds_{yy}}{Dt} = 2\mu\left(\dot{\varepsilon}_{yy} - \frac{1}{3}\frac{\dot{V}}{V}\right) - 2\mu\lambda s_{yy},$$

$$\frac{Ds_{xy}}{Dt} = 2\mu\dot{\varepsilon}_{xy} - 2\mu\lambda s_{xy}, \qquad \frac{Ds_{zz}}{Dt} = 2\mu\left(\dot{\varepsilon}_{zz} - \frac{1}{3}\frac{\dot{V}}{V}\right) - 2\mu\lambda s_{zz},$$

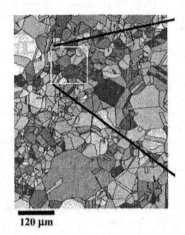

120 μm

Fig. 12. Map of representative mesovolume.

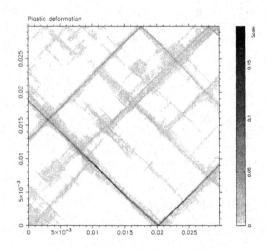

Plastic deformation

Fig. 13. Distribution of plastic strains

$$\sigma_{zz} = -P + s_{zz} = 0 \ - \qquad \text{for plane stress state,}$$
$$s_{zz} = -\left(s_{xx} + s_{yy}\right) - \qquad \text{for plane strain state,}$$

where $\dfrac{Ds_{ij}}{Dt} = \dot{s}_{ij} - \dot{\omega}_{ik}s_{kj} + s_{ik}\dot{\omega}_{kj}$ means Jaumann time derivative and $\dot{\lambda}$ is a scalar plastic

flow rate parameter defined with the von Mises yield condition $s_{xx}^2 + s_{yy}^2 + 2s_{xy}^2 + s_{zz}^2 = \dfrac{2}{3}Y_0^2$.

The dot over a parameter means a time derivative. Nomenclature: x, y are space coordinates, v_x is velocity in x direction, v_y is velocity in y direction, $\sigma_{xx}, \sigma_{yy}, \sigma_{xy}, \sigma_{zz}$ are stress tensor components, $s_{xx}, s_{yy}, s_{xy}, s_{zz}$ are stress deviators, $\varepsilon_{xx}, \varepsilon_{yy}, \varepsilon_{xy}, \varepsilon_{zz}$ are strain tensor components, P is hydrostatic pressure,

$V = \dfrac{\rho_0}{\rho}$ is relative volume,

where ρ_0 is reference density,
ρ is actual density,

K is bulk modulus, μ is shear modulus,
h is plate thickness,

Y_0 is yield strength.

For numerical solving this set of equations a computer program based on the finite-difference scheme known as Wilkins method was used [11-13]. An algorithm of splitting of grid nodes was applied to model fracture [14, 15]. Stresses of increased accuracy and smooth over the whole domain can be obtained from the nodal displacements and traction boundary conditions using T-functions as interpolators in connection with the Moving Least Squares (MLS) method [5]. We used full second order polynomials and thus for stress evaluation in each point of interest we need displacements in at least 5 nodal points for 2D problems and in 9 nodal points in 3D problem. Near the boundaries, additional conditions are obtained from boundary traction components in the closest boundary point, which increases the accuracy of computed stress field and decreases the number of necessary displacement nodal points in the domain of interest (DOI) for the interpolation. The displacements in internal nodal points, which are necessary for computation of the stress field inside the domain, are obtained from boundary tractions and displacements using the well known boundary integral presentation [1].

RESULTS AND DISCUSSION

The first example of simulation presents modeling of deformation of representative mesovolume of polycrystalline steel sample. A map of such mesovolume is submitted in fig. 12. There are more than 120 grains with average size about 3 microns in this volume and it can be called representative. In calculations for fragments of different color (grain) the yield strength was different (up to 30%). In uniaxial tension of such mesovolume the system of localized deformation bands take place with inclination of them about 45° to the axis of tension. In fig. 13,

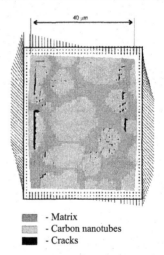

- Matrix
- Carbon nanotubes
- Cracks

Fig. 14. Velocity field in the representative mesovolume at fragmentation.

Fig. 15. Mesocrack formation in the composite material with plastic matrix and inclusions.

the greater values of intensity of plastic deformations $\varepsilon_i^{pl} = \sqrt{\dfrac{2}{3}\left(\varepsilon_{xx}^{pl\,2} + \varepsilon_{yy}^{pl\,2} + 2\varepsilon_{xx}^{pl\,2} + \varepsilon_{zz}^{pl\,2}\right)}$
correspond to greater intensity of coloring. The deformation in bands, covering on width about 0.5-0.8 microns, ranges up to 30% at integral deformation of a sample 0.7%. The system of bands of localized deformation breaks a sample into separate blocks (bulk structural elements) which move as units. This effect of material fragmentation during loading is distinctly seen in velocities field shown in fig. 14.

The last example of calculations shows deformation behavior up to failure of a ceramic composite with Al_2O_3 inclusions and Al matrix. Under complex loading (shear + tension) in a mesovolume of the composite a system of microcracks appear (fig. 15). In these calculations an algorithm of Lagrange grid nodes splitting was used. Here plastic matrix transmits loading to brittle inclusions, that crack while shear bands forming in matrix.

CONCLUSION

For a realistic simulation of the stability behavior of carbon nanotubes, the nonlinear intramolecular inter-actions between neighboring atoms have to be taken into account. A comparison shows the buckling sensitivity of different geometries. In order to reduce computational costs, it is necessary to develop suited homogenization techniques, so that shell elements can be applied.

It was shown the possibility of handling large gradients in displacement and stress fields by non-singular reciprocity based FEM technique using Kelvin functions with source points outside the domain. Trefftz polynomials of low order are used for the local interpolation of displacement in order to obtain the stresses inside the domain and on the domain and inclusion boundaries. The technique introduces considerable reduction of the problem. The numerical examples are shown for 2D problem; however the extension to 3D is straightforward.

Heterogeneity of stressed state is typical for deformation of mesovolumes of a structurally inhomogeneous material. This is due to stress concentrators of various nature and scale (interfaces of fragments of internal structure, feature of the shape etc.). In these conditions, the plastic deformation proceeds heterogeneously too. They arise in the region of stress concentration and in the least strength elements of structure. Then bands of localized shear are formed where the plastic deformations much exceed average deformations. In these bands significant change in values of shear and rotations making tensor of plastic distortion is marked, rotations being more sensitive to localization of deformation. The sign of rotations depends on orientation of a band concerning the axis of deformation. With localization of plastic deformation, material fragmentation occurs with formation of bulk structural elements, which move as units.

In the frameworks of classic elastic-plastic model by taking into account the heterogeneous inner structure of a material in explicit form and stress concentrators of various nature, it is possible to simulate numerically regions of localized plastic strain of meso scale that are observed in experiments.

ACKNOWLEDGEMENTS
The research works are being supported by the WELCH foundation scholarship, the American Vacuum Society – Dr. Frank Sheppard - administrator.

REFERENCES

[1]Dresselhaus M. S. and Avouris P. Introduction to carbon materials research. In M. S. Dresselhaus, G. Dresselhaus and P. Avouris (Eds.), Carbon Nanotubes, Topics in Applied Physics, 80 (2001) 1-9.

[2]Govindjee S. and Sackman J. L. On the use of continuum mechanics to estimate the properties of nanotubes. Solid State Communications, 110 (1999) 227-230.

[3]Lau K.T., Li H.L., Lim D.S. and Hui D. Recent Development and Future Research Trend on Nanotube/ Polymer Composite Materials. Annals of European Academy off Science Journal, 1 (2003) 314-330.

[4]Mayo S.L., Olafson B. D. and Goddard W. A. (III). DREIDING: A Generic Force Field for Molecular Simulations. J. Phys. Chem., 94 (1990) 8897-8909.

[5]Qian D., Wagner G. J., Liu W. K., Yu M. F. and Ruoff R. S. Mechanics of Carbon Nanotubes. In W. A. Goddard (III), D. W. Brenner, S. E. Lyshevski and G. J. Iafrate (Eds.), Handbook of Nanoscience, Engineering and Technology (2003) 19.1-19.63.

[6]Superfine R., Falvo M., Taylor R.M. (II) and Washburn S. Nanomanipulation: Buckling, transport, and rolling at the nanoscale. In W. A. Goddard (III), D. W. Brenner, S. E. Lyshevski and G. J. Iafrate, Handbook of Nanoscience, Engineering and Technology (2003) 13.1-13.20.

[7]Yakobson B.I. and Avouris, P. Mechanical Properties of Carbon Nanotubes. In M. S. Dresselhaus, G. Dresselhaus and P. Avouris Carbon Nanotubes, Topics in Applied Physics, 80 (2001) 287-327.

[8]A.Nadai, Theory of flow and fracture of solids, Vol. 1, 2nd ed., McGraw-Hill, New York (1950)

[9]V.E.Panin (ed.), Structural levels of plastic deformation and fracture, Nauka Publishing, Novosibirsk (1990) (in Russian)

[10]V.E.Panin (ed.), Physical Mesomechanics of Heterogeneous Media and Computer-Aided Design of Materials, Cambridge International Science Publishing, Cambridge (1998).

[11]M.L.Wilkins, Calculations of Elastic-Plastic Flow. in B.Alder, S.Fernbach and M.Rotenberg (eds), Methods in Computational Physics, Vol. 3, Academic Press, New York (1964) pp.211-263.

[12]M.L.Wilkins and M.W.Guinan, Plane Stress Calculations with a Two Dimensional Elastic-Plastic Computer Program, UCRL-77251 (1976)

[13]V.E. Panin, P.V. Makarov, I.Y. Smolin, O.I. Cherepanov, V.N. Demidov, M.M. Nemirovich-Danchenko, Methodology of computer-aided design of materials with specified strength characteristics, in: V.E. Panin (Ed.), Physical Mesomechanics of Heterogeneous Media and Computer-Aided Design of Materials, Cambridge International Science Publishing, Cambridge, 1998, pp. 199–249.

[14]V.A.Gridneva, M.M.Nemirovich-Danchenko, Method of grid nodes splitting for calculations of solids fracture, Dep. in VINITI, Tomsk, No. 3258 (1983) (in Russian)

[15]J. Balaš, J. Sládek, V. Sládek, Stress Analysis by Boundary Element Methods, Elsevier, Amsterdam (1989).

[16]N. I. Muskhelishvili, Some Basic Problems of Mathematical Theory of Elasticity, (in Russian), Nauka, Moscow (1966).

[17]B. I. Blokh, Theory of Elasticity, (in Russian), Univ. Press Kharkov (1964).

[18]V. D. Kupradze, Potential Methods in Theory of Elasticity, (in Russian), Nauka, Moscow (1963).

[19]V. Kompiš, Selected Topics in Boundary Integral Formulations fo Solids and Fluids, CISM Courses and Lectures No.433, Springer – Verlag, Wien, in print.

Author Index

Keyword Index